GRAPHIC GUIDE TO
Site Construction

OVER 325 DETAILS FOR BUILDERS AND DESIGNERS

GRAPHIC GUIDE TO
Site Construction

OVER 325 DETAILS FOR BUILDERS AND DESIGNERS

ROB THALLON and STAN JONES

The Taunton Press

The Taunton Press
Inspiration for hands-on living®

The Taunton Press, Inc., 63 South Main Street, PO Box 5506, Newtown, CT 06470-5506
e-mail: tp@taunton.com

Distributed by Publishers Group West

EDITOR: Stefanie Ramp
INTERIOR DESIGNER: Lynne Phillips
LAYOUT: Cathy Cassidy
ILLUSTRATOR: Rob Thallon's drawings rendered by Vince Babak
FRONT COVER PHOTOGRAPHER: ©Lee Anne White

For Pros By Pros® is a trademark of The Taunton Press, Inc.,
registered in the U.S. Patent and Trademark Office.

LIBRARY OF CONGRESS CATALOGING-IN-PUBLICATION DATA:

Thallon, Rob.
    Graphic guide to site construction : over 325 details for builders and
designers / Rob Thallon and Stan Jones.
        p. cm. -- (For pros, by pros)
    ISBN 1-56158-549-1
    1. Landscape construction. 2. Building sites. 3. Landscape
architectural drawings. I. Jones, Stan. II. Title. III. Series.

    TH380 .T45 2002
        690'.89--dc21

                                2002152505

Printed in United States of America
10 9 8 7 6 5 4 3 2 1

TO DEE, CARTER, AND CLAIRE

TO LAURA, JORDAN, RYAN, AND KYLER

ACKNOWLEDGMENTS

This book has been enriched immeasurably by the contributions of consultants throughout the country. The following design and construction professionals have reviewed a draft of each chapter. Their participation has made the book more comprehensive and the process of writing it more enjoyable.

Richard Anderson
Sarasota, Florida

Joe Ewan
Tempe, Arizona

Daniel Winterbottom
Seattle, Washington

Also, we gratefully acknowledge the invaluable contributions of the following people:

Steve Culpepper, for the original idea for this volume and for his shepherding of that idea into reality;

Jennifer Renjilian, our original editor, for getting the project off the ground;

Stefanie Ramp, our editor, for gracious management and astute tuning of the writing;

Scott Wilkinson and David Bloom, illustrators, for their research and help with development of the drawings;

All of the other professionals whose contributions, large and small, have added to the quality of this volume;

Our colleagues in the departments of Architecture and Landscape Architecture at the University of Oregon, for their suggestions and words of encouragement throughout the development of this project;

And finally, our families, whose patience with our absurd schedules have allowed us the time to focus on this project.

CONTENTS

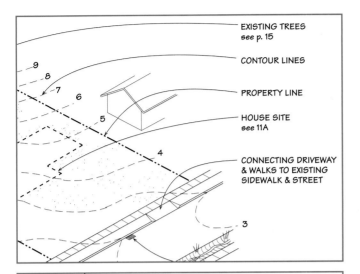

EXISTING TREES
see p. 15

CONTOUR LINES

PROPERTY LINE

HOUSE SITE
see 11A

CONNECTING DRIVEWAY
& WALKS TO EXISTING
SIDEWALK & STREET

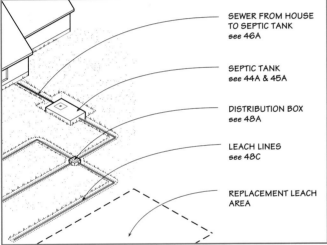

SEWER FROM HOUSE
TO SEPTIC TANK
see 46A

SEPTIC TANK
see 44A & 45A

DISTRIBUTION BOX
see 48A

LEACH LINES
see 48C

REPLACEMENT LEACH
AREA

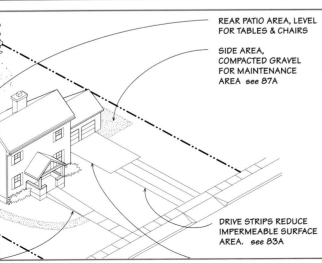

REAR PATIO AREA, LEVEL
FOR TABLES & CHAIRS

SIDE AREA,
COMPACTED GRAVEL
FOR MAINTENANCE
AREA see 87A

DRIVE STRIPS REDUCE
IMPERMEABLE SURFACE
AREA. see 83A

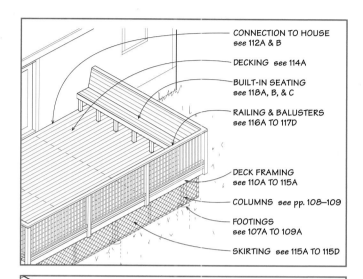

CONNECTION TO HOUSE
see 112A & B

DECKING see 114A

BUILT-IN SEATING
see 118A, B, & C

RAILING & BALUSTERS
see 116A TO 117D

DECK FRAMING
see 110A TO 115A

COLUMNS see pp. 108–109

FOOTINGS
see 107A TO 109A

SKIRTING see 115A TO 115D

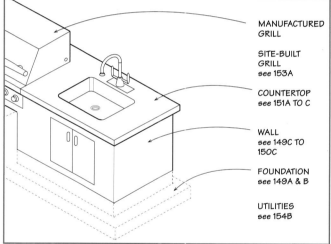

MANUFACTURED
GRILL

SITE-BUILT
GRILL
see 153A

COUNTERTOP
see 151A TO C

WALL
see 149C TO
150C

FOUNDATION
see 149A & B

UTILITIES
see 154B

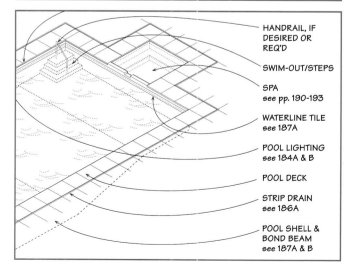

HANDRAIL, IF
DESIRED OR
REQ'D

SWIM-OUT/STEPS

SPA
see pp. 190-193

WATERLINE TILE
see 187A

POOL LIGHTING
see 184A & B

POOL DECK

STRIP DRAIN
see 186A

POOL SHELL &
BOND BEAM
see 187A & B

INTRODUCTION

Every year, more than a million new residences are constructed in North America, and with them comes the development of yards and gardens that allow the occupants of these new dwellings to move their daily lives outside. In addition to new gardens, hundreds of thousands of existing gardens are improved and remodeled each year.

A well-planned and well-constructed yard extends the living space of a house in innumerable ways. In southern climates, where the weather is agreeable more often than not, the convenience of usable outdoor spaces in the form of porches, terraces, and lawns is taken for granted. And in northern climates, where enjoyable weather is less frequent, the use of outdoor spaces is more precious and their importance elevated. Even when the site is used for neither social activities nor play, there are usually a number of exterior items to be designed and built that make a residence complete.

The development of a yard is not a simple thing. Even the most basic outdoor space with a deck, a fence, and a lawn require knowledge of materials, structure, drainage, soils, planting, and detailing for the weather. And more developed gardens that contain lighting, pathways, retaining walls, irrigation, and other complex components require deeper knowledge. Most people are not familiar enough with these systems to start building them without some research. So it occurred to us that this basic information, collected into one volume, would be a useful tool for builders, designers, homeowners, and students alike.

This volume complements two other Graphic Guides—*Graphic Guide to Frame Construction*, a reference for the construction of wood frame buildings; and *Graphic Guide to Interior Details*, a reference for finishing the interior of these (and other) buildings. This *Graphic Guide* provides the necessary tools to complete the site construction so that, with all three guides, details are provided for an entire residential building project.

THE SCOPE OF THE BOOK

Unlike the other *Graphic Guides*, the scope of this book is limited to residential projects. This was necessary because sitework changes dramatically when it moves from the domestic to the public scale. But within that single limitation, the intention in selecting the material for this guide was to include information typical to all sites. Thus, this volume will serve as a reference for all residential-scale projects in North America, including both new construction and remodeling. It covers all aspects of site development up until the time that plants go in the ground.

The reason that plants are excluded has to do with the complexity of the subject. There are so many different climatic zones and so many plant species, each with unique requirements, that even a condensed guide to planting scarcely can be forced into one volume. Add to that the idiosyncratic nature of people's relationship to plants, and the task becomes even more formidable. But the groundwork of site construction—the grading, drainage, utilities, paving, and elements such as fences and decks that form the practical and spatial framework for the planting—is more or less universal. Principles and details employed in New Orleans will translate directly (except for frostline) to Minneapolis.

Although all topics are covered, not all are treated with the same level of detail. The intent is to provide details for the construction of elements that contractors or homeowners are likely to build themselves, but only an overview of elements usually subcontracted to a specialist. Thus, the construction of fences and walkways are covered in detail, while the design and construction of septic systems and swimming pools are explained only conceptually. In all cases, references for further research are cited.

FOCUS ON SUSTAINABILITY

Although the details in this book have been selected primarily on the basis of their widespread use, there also has been an attempt to include procedures and details that contribute to environmental responsibility. Partly, this involves selecting practices for their efficiency of material use, their low-energy use, their use of renewable and recyclable resources, and their nontoxic components. And partly, it involves the inclusion of practices and materials that are durable to avoid the consumption of resources (both environmental and economic) required by frequent rebuilding.

Sustainability, however, is a complex concept that involves judgment and the balancing of numerous factors for an elusive long-term goal. For example, is it better to use preservative-treated wood for increased longevity or untreated wood for decreased toxicity? This type of question can best be answered on a project-by-project basis because it is subject to a variety of site-specific variables such as annual rainfall and intended use. We have attempted to include practices and details that should be considered when weighing these variables.

ON CODES

Every effort has been made to ensure that the statements and details included in this book conform to building codes. Codes vary, however, so local codes and building departments always should be consulted to verify compliance.

HOW THE BOOK WORKS

The book's six chapters follow the approximate order of construction, starting with site grading and working up to the finished landscape—ready for planting. Most chapters cover more than one major subject and are divided into subsections, also roughly ordered according to the sequence of construction. Chapter titles and major subsections are called out at the top of each page for easy reference.

The pages are numbered at the top outside corner, and all the drawings are lettered. With this system, all the drawings may be cross-referenced. The callout "see 119B," for example, refers to drawing B on page 119. (See the example on facing page.) Drawings may be referenced from the text or from other drawings.

Any notes included in a detail are intended to describe its most important features. By describing the relationship of one element to another, the notes sometimes go a little further than merely naming the feature. Material symbols are described on page 195. Abbreviations are spelled out on page 197.

A FINAL NOTE

Our intention in writing this book has been to assist designers and builders who are attempting to make practical and beautiful outdoor spaces that endure. With the drawings and text, we have tried to describe the relationship among all the common components of such spaces. Alternative approaches to popular practices have been included as well.

We have relied primarily on our own experiences but have also drawn significantly on the accounts of others. In order to build upon this endeavor, we encourage you, the reader, to inform us of your own observations and/or to make critical comments. Please send them to us care of Books Department, The Taunton Press, P.O. Box 5506, Newtown, CT 06470-5506.

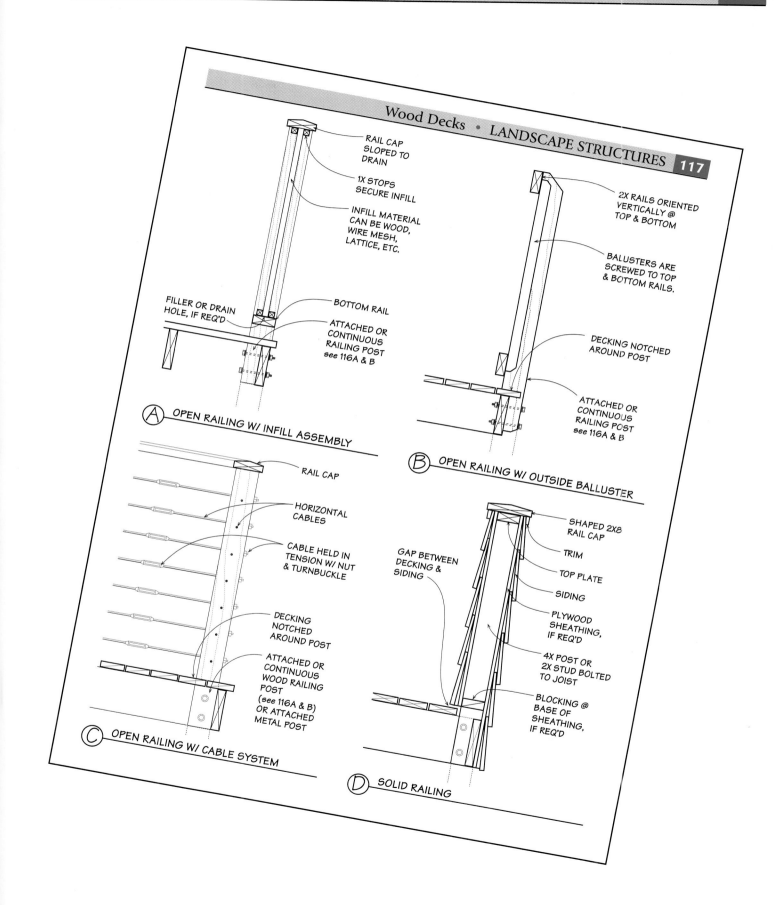

Wood Decks • LANDSCAPE STRUCTURES 117

RAIL CAP
SLOPED TO
DRAIN

1X STOPS
SECURE INFILL

INFILL MATERIAL
CAN BE WOOD,
WIRE MESH,
LATTICE, ETC.

FILLER OR DRAIN
HOLE, IF REQ'D

BOTTOM RAIL

ATTACHED OR
CONTINUOUS
RAILING POST
see 116A & B

Ⓐ OPEN RAILING W/ INFILL ASSEMBLY

2X RAILS ORIENTED
VERTICALLY @
TOP & BOTTOM

BALUSTERS ARE
SCREWED TO TOP
& BOTTOM RAILS.

DECKING NOTCHED
AROUND POST

ATTACHED OR
CONTINUOUS
RAILING POST
see 116A & B

Ⓑ OPEN RAILING W/ OUTSIDE BALLUSTER

RAIL CAP

HORIZONTAL
CABLES

CABLE HELD IN
TENSION W/ NUT
& TURNBUCKLE

DECKING
NOTCHED
AROUND POST

ATTACHED OR
CONTINUOUS
WOOD RAILING
POST
(see 116A & B)
OR ATTACHED
METAL POST

Ⓒ OPEN RAILING W/ CABLE SYSTEM

GAP BETWEEN
DECKING &
SIDING

SHAPED 2X8
RAIL CAP

TRIM

TOP PLATE

SIDING

PLYWOOD
SHEATHING,
IF REQ'D

4X POST OR
2X STUD BOLTED
TO JOIST

BLOCKING @
BASE OF
SHEATHING,
IF REQ'D

Ⓓ SOLID RAILING

CHAPTER 1

SITE GRADING AND DRAINAGE

Grading is the act of shaping the land. This includes large-scale earth-moving for the construction of a house or other building as well as smaller scale efforts, such as collecting or directing water, creating mounds of earth for raised planting areas, or making an open area suitable for play. While ensuring proper drainage and maximum soil stability are two of the most important aspects of grading any site, many other critical issues may arise that will affect what shape the landscape should ultimately possess.

There are two basic methods employed to shape a landscape—cutting and filling. A "cut" refers to an area where soil has been removed, while a "fill" indicates an area where soil has been added. From an economic and often an ecological standpoint, attempting to balance cut and fill is the desired goal. That means that for all of the soil dug up on a site, there is a corresponding location for it to be placed in the final site design.

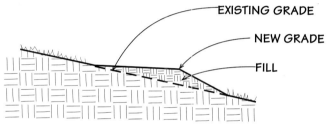

TERRACE IN FILL (SECTION)

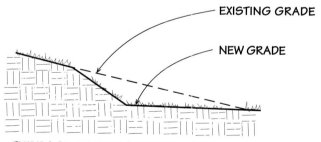

TERRACE IN CUT (SECTION)

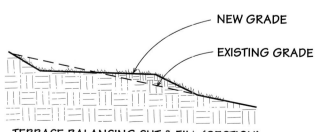

TERRACE BALANCING CUT & FILL (SECTION)
NOTE: THIS METHOD SAVES COST IN TIME,
MATERIALS, & TRANSPORTATION.

READING GRADING PLANS

Grading plans account for many of the drawings in this chapter, and they depict three-dimensional landforms through the use of contour lines. Contour lines are graphic lines that denote a constant elevation above a given elevation at a predefined spot, often referred to as a datum point. They are the best way in which to illustrate in two dimensions a landscape that exists in three dimensions—length, width, and height.

Graphically, contours in their preconstruction or existing form are depicted most often as dashed lines, while contours that are changed, or will be changed during construction, are depicted as solid lines. In many instances, parts of contours may remain unchanged (shown as dashed lines), while other parts of the same contour will be altered to give the landscape its desired shape (depicted with a solid line). Typically, the old location of the contour that has been changed is left showing so that the builder will know how much earth will be removed or brought to the site. Sometimes, for purposes of clarity, existing contour lines that are to be changed will have their altered portion faded out.

To understand how contours relate to landform, consider that water always flows downhill perpendicular to a contour line and that the steepest route of travel is the shortest distance between two contours. As contour lines come closer together, the landform gets steeper; as they move farther apart, the landform becomes flatter.

Contours are labeled with a number that denotes an elevation relative to the datum point; in the United States, it's typically measured in feet, or portions of feet. In other countries, meters are used. In all instances, it is important to remember that contours depict both vertical and horizontal distance—vertical through the numerical elevations assigned to the lines (the vertical change from one contour to the next is defined as the contour interval), and horizontal through the measurable distance between contours as depicted in the drawing (assuming that the drawing is done to scale). Typical contour intervals in topographic plans are 1 ft., 2 ft., 5 ft., and 10 ft.; the smaller the contour interval, the more accurate a depiction of the landform a plan will give you.

Other drawings in this chapter, such as sections and detail drawings, are for purposes of illustrating principles and reminding the designer and the builder to consider their use carefully. These drawings, therefore, should be used only as a reference.

GRADING AND DRAINAGE BASICS

Grading and drainage are often referred to as two sides of the same coin: Whenever the shape of the land is altered, the movement of water is altered as well. Therefore, thinking about how water moves in and through a site prior to construction is critical in determining how the site plan is going to create new drainage patterns.

Surface-drainage patterns fall into two categories: water collection and water dispersion (also called sheet flow); most sites incorporate elements of both. Water-collection strategies direct water toward a central collecting point. Sometimes this conveyance is achieved in the form of a swale (a depression in the landscape similar to a small stream channel in form, though much more shallow, and in many instances, nearly invisible). Water also may be directed into a drainage structure, such as a catch basin or drain inlet, that feeds water into an underground pipe. Often, a site will incorporate elements of above-ground and below-ground systems, taking advantage of the positive aspects of each. (These will be discussed later in the chapter.)

Water-dispersion techniques work best in well- or fairly well-drained soils where a portion of the water can percolate, or enter the subsurface water flow. Sheet flows also can work in areas with poorly drained soils, but the landscape will need a steeper slope. Ultimately, unless the soils are extremely porous and/or well-drained, areas that are sheet drained will need to feed into a retention pond or a drainage structure that will convey the water away from the site.

The primary benefit of sheet-flow drainage is that it allows an area to retain maximum functionality, interfering with the least number of possible uses. In the case of paved areas, for instance, it accommodates court-related games, patio furniture, and easy pedestrian access.

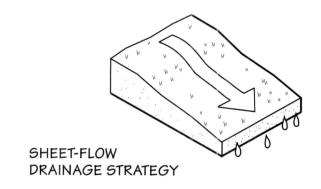

SHEET-FLOW
DRAINAGE STRATEGY

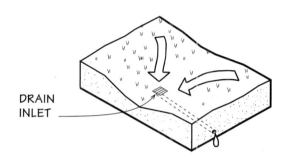

DRAIN
INLET

WATER-COLLECTION
DRAINAGE STRATEGY

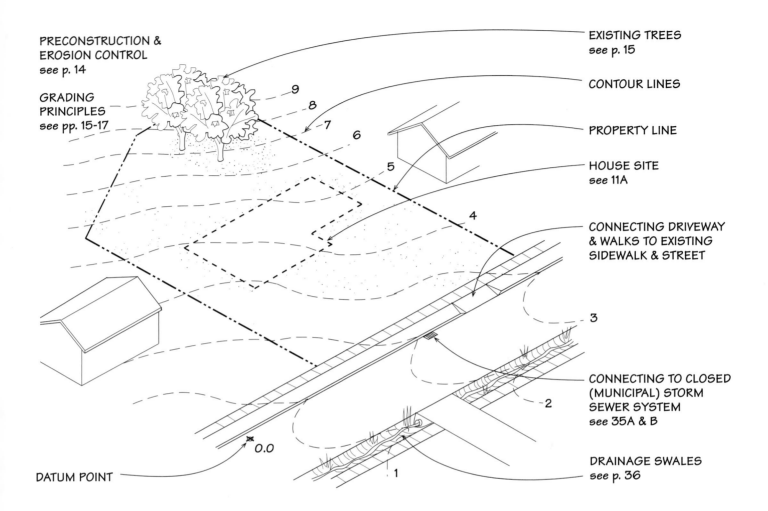

PRECONSTRUCTION &
EROSION CONTROL
see p. 14

GRADING
PRINCIPLES
see pp. 15-17

EXISTING TREES
see p. 15

CONTOUR LINES

PROPERTY LINE

HOUSE SITE
see 11A

CONNECTING DRIVEWAY
& WALKS TO EXISTING
SIDEWALK & STREET

CONNECTING TO CLOSED
(MUNICIPAL) STORM
SEWER SYSTEM
see 35A & B

DRAINAGE SWALES
see p. 36

DATUM POINT

0.0

SITE CONTEXT–The topography of a given site is a critical consideration when designing and/or constructing anything that relates to, sits on, or connects to the ground. Recognizing existing patterns and conditions can and should affect decisions about how the land will be shaped, what type of walls and/or foundations (if any) you might choose to construct, and how the overall site plan might be organized.

A primary concern is how surface and subsurface drainage is flowing through and around your site prior to construction. As suggested earlier, however, there are many other aspects of your site's design that can and should direct how the earth is shaped. These include functional goals, such as slope stability, on-site drainage, and the preservation of existing vegetation, as well as aesthetic goals or the desire to accommodate specific activities, such as gardening, picnick-

ing, or soccer. There even may be issues specific to your site that will inform its design, such as matching existing grades on adjacent properties or along streets and/or sidewalks that may impact human or vehicular accessibility. Finally, proximity to sensitive natural areas, such as creeks or woodlands, may require consideration in both the design and construction process to minimize any ecological disturbance.

As part of assessing the site conditions, a good survey of existing elevations both on a site and of important elements and/or systems around a site is a crucial first step in developing a plan that will meet the homeowner's needs and expectations. Knowing the elevation of the street, for example, can be critical in determining the Finished Floor Elevation (FFE) of the house or structure being designed.

The FFE is most typically defined as the elevation of the entry level floor inside the threshold of each door. If the FFE is set too low, the finished grades adjacent to the foundation will make it difficult to drain water away from the structure. The addition of drainage structures or other potentially costly steps could be necessary to help protect the foundation's integrity and prevent an overly soggy surrounding landscape. If the FFE is set too high, steps or walls could be needed.

Significant elevation changes can have an enormous impact on the accessibility of a home and landscape (for many, the addition of a single step can eliminate access to areas within the landscape or house itself). When determining the relationship of the inside elevation to the outside of the home, take into consideration the impact that a change in vertical distance can have on the day-to-day life of a home, from lugging groceries to moving furniture to providing access for strollers, walkers, or wheelchairs. Careful setting of the FFE is a very important step in developing an overall strategy for the grading of the site and should be done with a "big picture" approach to site design (see 11A).

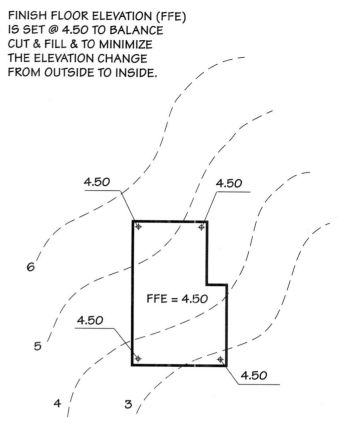

FINISH FLOOR ELEVATION (FFE) IS SET @ 4.50 TO BALANCE CUT & FILL & TO MINIMIZE THE ELEVATION CHANGE FROM OUTSIDE TO INSIDE.

4.50   4.50

6

FFE = 4.50

4.50

5

4.50

4   3

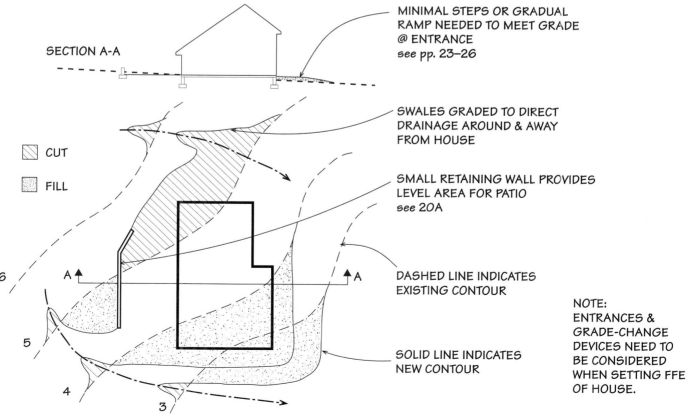

SECTION A-A

MINIMAL STEPS OR GRADUAL RAMP NEEDED TO MEET GRADE @ ENTRANCE
see pp. 23–26

SWALES GRADED TO DIRECT DRAINAGE AROUND & AWAY FROM HOUSE

SMALL RETAINING WALL PROVIDES LEVEL AREA FOR PATIO
see 20A

CUT

FILL

DASHED LINE INDICATES EXISTING CONTOUR

6

A ▲   ▲ A

NOTE:
ENTRANCES & GRADE-CHANGE DEVICES NEED TO BE CONSIDERED WHEN SETTING FFE OF HOUSE.

5

SOLID LINE INDICATES NEW CONTOUR

4

3

Ⓐ BASIC GRADING FOR A HOUSE SITE

Setting the FFE of a structure is also somewhat dependent on what type of foundation is being planned for the building in question (see 12A). The type of foundation, be it a slab, stem wall, basement, or pier/post foundation, will impact the nature of how drainage can be addressed. In all instances, drainage away from the foundation wall and structure is desirable, since wide fluctuations in soil moisture can lead to problems with the soil and the foundation itself.

Other elevations that may be critical to note are those either inside the site or on the perimeter that cannot change. Elevations that cannot change may include existing structures on other properties or streams that run adjacent to or through the site. Other elevations that you may not want to change, such as the elevations around the drip lines of existing trees that you wish to save, also should be noted.

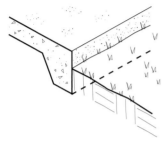

TURNED-DOWN SLAB FOUNDATIONS OFFER THE LEAST FLEXIBILITY FOR GRADING AROUND THE HOUSE.

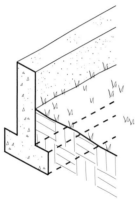

WALL FOUNDATIONS PROVIDE MORE FLEXIBILITY FOR GRADING AROUND THE HOUSE DUE TO INCREASED HEIGHT OF CONCRETE POUR.

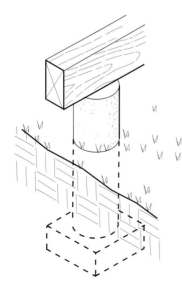

PIER FOUNDATIONS CAN PROVIDE THE MOST FLEXIBILITY FOR GRADING AROUND THE HOUSE. THEY ALSO CAN BE USEFUL ON VERY STEEPLY SLOPING SITES BY REDUCING CUT & FILL & ALLOWING SHEET FLOW UNDERNEATH THE STRUCTURE.

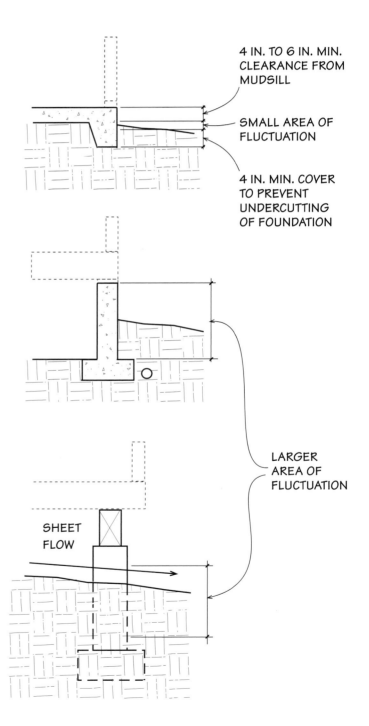

4 IN. TO 6 IN. MIN. CLEARANCE FROM MUDSILL

SMALL AREA OF FLUCTUATION

4 IN. MIN. COVER TO PREVENT UNDERCUTTING OF FOUNDATION

LARGER AREA OF FLUCTUATION

SHEET FLOW

SOILS—The nature and type of soil that exists on a site is important to understand, as it impacts much of the early construction on a project. Preliminary grading and excavation work should take into account the stability and strength of the soils on-site, and it should preserve as much of the fertile topsoil as possible. In areas where compaction of topsoil is likely or where structures such as buildings, driveways, and patios are planned, the topsoil should be excavated and stockpiled on-site for reuse. This leaves behind what is called the subsoil, which has less decomposing organic matter in it and, therefore, is (typically) more structurally sound.

For construction purposes, soils with high clay content that will "shrink and swell" (i.e., expand with the addition of moisture and contract with the loss of moisture, as might happen during dry spells in the summer) should be avoided if possible for anything requiring a foundation, such as a building, wall, or driveway. In places where clay is completely prevalent throughout a site, it is recommended to excavate as much of it as possible in the areas where foundations and other structural, built elements are being constructed. It should be replaced with a layer of very well-drained river rock or engineered fill (a mixture of rock and gravels, sometimes accompanied by layers of geotextile fabric to reduce settling) that is compacted in place by a mechanical compactor.

Similarly, in areas with soils composed of highly organic, decomposable materials, or in areas where previous uses may have left buried areas of rubble or fill, take care to ensure that all unstable matter has been removed and that a solid base of well-drained fill, preferably rock, has been installed in its place prior to construction.

For a detailed assessment of the structural capabilities of the soil on a specific site, consult a soils or structural engineer, or check with your local planning and permitting agencies, as they may have standards based upon prevailing local conditions.

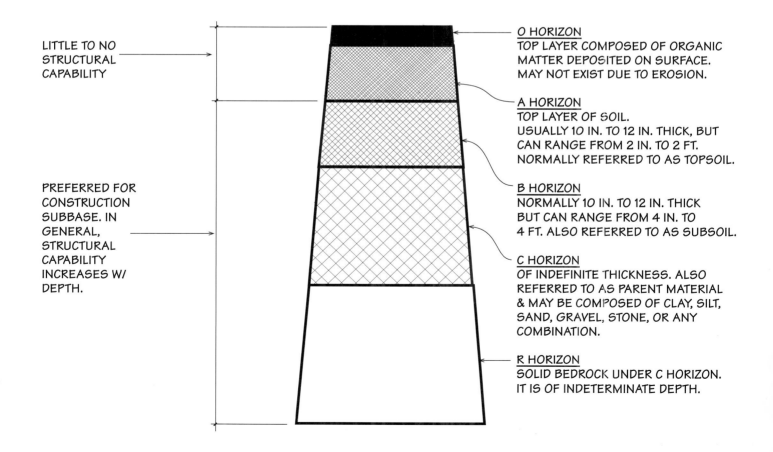

LITTLE TO NO STRUCTURAL CAPABILITY

PREFERRED FOR CONSTRUCTION SUBBASE. IN GENERAL, STRUCTURAL CAPABILITY INCREASES W/ DEPTH.

O HORIZON
TOP LAYER COMPOSED OF ORGANIC MATTER DEPOSITED ON SURFACE. MAY NOT EXIST DUE TO EROSION.

A HORIZON
TOP LAYER OF SOIL. USUALLY 10 IN. TO 12 IN. THICK, BUT CAN RANGE FROM 2 IN. TO 2 FT. NORMALLY REFERRED TO AS TOPSOIL.

B HORIZON
NORMALLY 10 IN. TO 12 IN. THICK BUT CAN RANGE FROM 4 IN. TO 4 FT. ALSO REFERRED TO AS SUBSOIL.

C HORIZON
OF INDEFINITE THICKNESS. ALSO REFERRED TO AS PARENT MATERIAL & MAY BE COMPOSED OF CLAY, SILT, SAND, GRAVEL, STONE, OR ANY COMBINATION.

R HORIZON
SOLID BEDROCK UNDER C HORIZON. IT IS OF INDETERMINATE DEPTH.

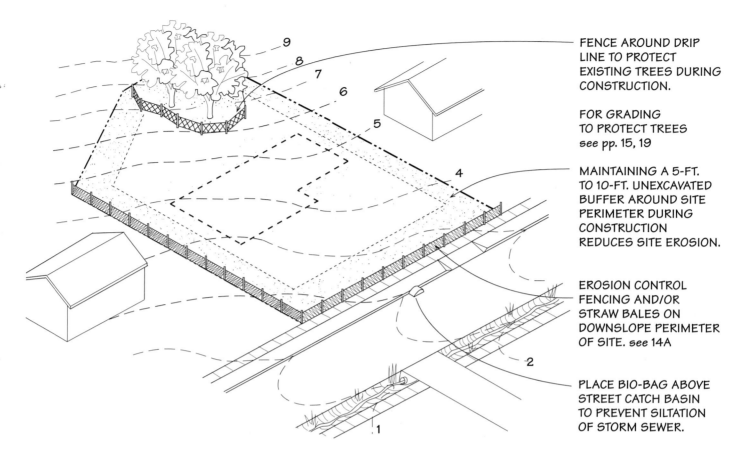

FENCE AROUND DRIP LINE TO PROTECT EXISTING TREES DURING CONSTRUCTION.

FOR GRADING TO PROTECT TREES *see pp. 15, 19*

MAINTAINING A 5-FT. TO 10-FT. UNEXCAVATED BUFFER AROUND SITE PERIMETER DURING CONSTRUCTION REDUCES SITE EROSION.

EROSION CONTROL FENCING AND/OR STRAW BALES ON DOWNSLOPE PERIMETER OF SITE. *see 14A*

PLACE BIO-BAG ABOVE STREET CATCH BASIN TO PREVENT SILTATION OF STORM SEWER.

GRADING AND DRAINAGE PRINCIPLES

When shaping the land on a given site, there are several basic principles to keep in mind, regardless of how the landscape on your site is designed. While not an exhaustive list, these are critical items to keep in mind as your project progresses.

• **Erect preventive erosion-control measures prior to construction.** Legislation such as the Clean Water Act and other federal, state, and local policies are dictating that builders be much more careful about how their practices impact local streams and water quality on- and off-site. Therefore, prior to any construction activity on your site, take measures to prevent soil erosion. Typical steps could include erecting a perimeter barrier (either with fabric or with hay or straw bales) or installing "bio-bags" (mesh bags typically filled with wood chips or some other filtration material) to prevent soil from entering storm-sewer systems. Check with your local city permitting agency to determine what the local requirements are.

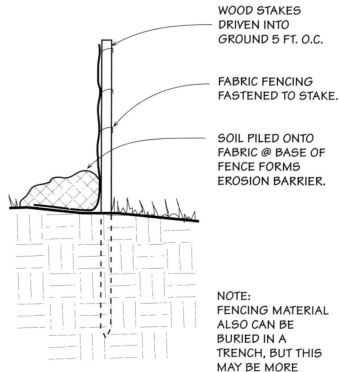

WOOD STAKES DRIVEN INTO GROUND 5 FT. O.C.

FABRIC FENCING FASTENED TO STAKE.

SOIL PILED ONTO FABRIC @ BASE OF FENCE FORMS EROSION BARRIER.

NOTE: FENCING MATERIAL ALSO CAN BE BURIED IN A TRENCH, BUT THIS MAY BE MORE DISRUPTIVE.

 A　EROSION-CONTROL FENCING

• **Preserve existing grades around trees you wish to save.** Trees or other existing vegetation to be preserved should be fenced off with a temporary but sturdy fence (chain-link panels or a well-staked orange construction fence, for example) around drip lines prior to construction. This will prevent accidental soil compaction by workers and equipment under their canopies, and reinforce the need to maintain the preexisting grade around the area. Soil compaction is a primary culprit in the premature death of trees on construction sites because it decreases the soil's ability to allow air and water exchange in the root zone of the tree.

THE DRIP LINE OF A TREE, X, (OR THE BRANCHING PERIMETER) GENERALLY CORRESPONDS TO THE SPREAD OF ITS ROOTS

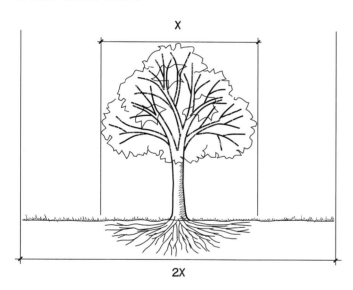

HOWEVER, MANY SPECIES HAVE A CRITICAL ROOT ZONE OF UP TO 200% OF THE DRIP LINE.

Similarly, altering the grade underneath a tree also can severely impact the ability of the roots to perform up to a level that a healthy plant requires. Current research has shown that preserving the existing grades and drainage patterns in an area roughly one-and-a-half to two times the diameter of the drip line of a tree provides the best opportunity for that tree to live and thrive beyond the period of construction. Changing the natural drainage pattern on a site, especially increasing or decreasing the amount of storm or irrigation runoff into the area around the drip line, can also damage a tree.

• **Drain water away from the house or structure.** The landscape surrounding a house or structure should have a minimum of 5 ft. of "positive," or downhill, drainage away from the base of the foundation if at all possible. In some instances, the use of drainage structures such as perforated pipe or catch basins might need to be used to help facilitate drainage in difficult situations. Maintaining positive drainage away from the foundation is critical to the health and longevity of the foundation itself, as well as to the structure.

• **Avoid draining onto adjacent properties.** Nearly all jurisdictions place restrictions on how much (if any) water can drain from one property onto another. Generally, a good measure to work toward is maintaining or slightly decreasing the amount of water draining onto an adjacent property. Avoid concentrating flows of water near property lines, and where concentrated flows are created, disperse the concentration as much as possible prior to its leaving the site. Many cities and jurisdictions also will require a 5-ft. to 10-ft. "buffer zone" around the edge of a property that is set aside for the purposes of "merging" the landforms of the two adjacent pieces of property. There are exceptions in extreme conditions (such as building in very hilly topography) or in places where local jurisdictions will allow the construction of retaining walls along the property line.

• **Drain away from high use or traffic areas.** As the landscape takes form, take care to match the desired usage of the landscape with the drainage design for a site. High-activity areas, such as a lawn where soccer or other games will be played, should be very gently sloped, with no areas of concentrated water flow.

Similarly, paved areas where patio furniture is to be set up should be drained using a sheet flow, with slopes not exceeding 2 percent (1 ft. of fall for every 50 ft. of horizontal distance, or run; see chart below). Edges between lawns and planted areas can be good areas to direct water flow, and possibly even concentrate it into a swale that will channel water away from higher use zones. Drain away from areas where heavy pedestrian or vehicle traffic is anticipated.

Element	Max	Min	Preferred		Notes
Driveways					0.5% is minimum for all impervious surfaces (or 1/2 in. per 10 ft.)
Longitudinal (Direction of Travel)	20%	0.5%	1–10%	5%	
Cross Slope (Side to Side)	10%	0.5%	1–3%	2%	
Walks					ADA requirement is a maximum of 8% with a landing every 30'
Longitudinal	10%	0.5%	1–5%	3%	
Cross Slope	4%	1%	2%	2%	(or 9 1/2 in. per 10 ft.)
Patios					
Concrete	2%	0.5%	1%		
Flagstone or Brick Pavers	2%	0.75%	1%	1%	
Greenspace					
Lawn Areas	25%	2%	Varies	25% 30%	Difficult to mow slopes greater than 25%
Planted Slopes	30%	2%	Varies		Varies with soil & plant type

Note: 1%= 1.25 in. per 10 ft. of horizontal surface

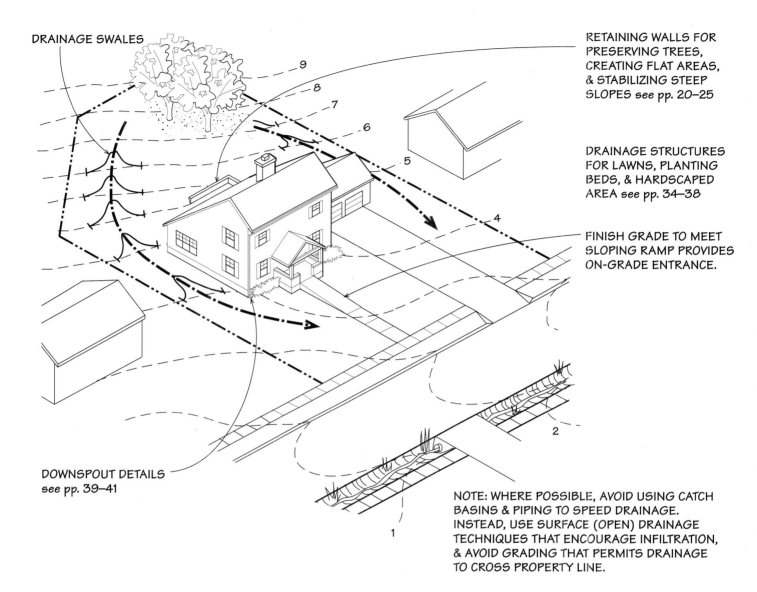

DRAINAGE SWALES

RETAINING WALLS FOR PRESERVING TREES, CREATING FLAT AREAS, & STABILIZING STEEP SLOPES see pp. 20–25

DRAINAGE STRUCTURES FOR LAWNS, PLANTING BEDS, & HARDSCAPED AREA see pp. 34–38

FINISH GRADE TO MEET SLOPING RAMP PROVIDES ON-GRADE ENTRANCE.

DOWNSPOUT DETAILS see pp. 39–41

NOTE: WHERE POSSIBLE, AVOID USING CATCH BASINS & PIPING TO SPEED DRAINAGE. INSTEAD, USE SURFACE (OPEN) DRAINAGE TECHNIQUES THAT ENCOURAGE INFILTRATION, & AVOID GRADING THAT PERMITS DRAINAGE TO CROSS PROPERTY LINE.

- **If possible, avoid using catch basins and piping to speed drainage.** Opt for surface, or open, drainage techniques instead. Using piped systems that involve catch basins and drains is necessary in many situations, but they are more costly than open surface techniques, such as sheet drainage or swales. Drains also can negatively impact local ecology, specifically the health of local streams, since they often outlet into local water bodies. Overland flow techniques slow runoff, allowing a significant amount of water to percolate into the subsurface soil, which mimics the way in which water naturally moves through a site prior to construction. Finally, subsurface drainage systems require more consistent monitoring and a higher level of maintenance in many cases compared with open drainage systems.

FUNCTION, AESTHETICS, AND COST

While deciding on what uses and contextual issues will drive the design of your landscape, it is important to consider the many grading and drainage options that are available to assist you in realizing your goal. When shaping the land, the primary, and in many cases, cheapest, element to consider is the on-site soil itself. Avoiding soil removal from the site, while attempting to use what exists without reliance on "built" elements (retaining walls, steps, and catch basins) can help keep costs down significantly.

There are, however, other kinds of costs associated with using slopes instead of walls, or swales instead of catch basins. These costs are most typically defined in terms of on-site landscape space dedicated to the

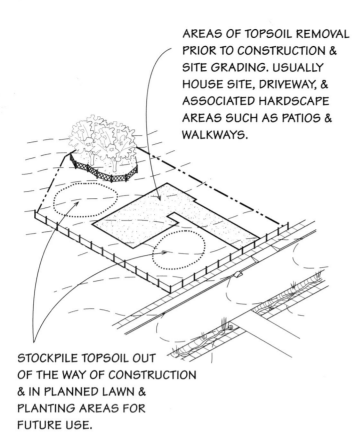

AREAS OF TOPSOIL REMOVAL PRIOR TO CONSTRUCTION & SITE GRADING. USUALLY HOUSE SITE, DRIVEWAY, & ASSOCIATED HARDSCAPE AREAS SUCH AS PATIOS & WALKWAYS.

STOCKPILE TOPSOIL OUT OF THE WAY OF CONSTRUCTION & IN PLANNED LAWN & PLANTING AREAS FOR FUTURE USE.

 MANAGING TOPSOIL ON BUILDING SITE

task of making up for vertical elevation change or to handling storm-water runoff. Planted slopes that make up the grade difference between home sites can be beautiful landscape amenities, but they are also space intensive and take up area that might otherwise be used for activities such as playing ball or entertaining.

Similarly, a swale transporting water can be designed to be nearly invisible, or it can be a highly attractive landscape element. In both cases, they are wonderful tools that can benefit the environment, allowing percolation and delaying runoff into overtaxed urban streams and rivers. They can, however, impact the usability of certain areas if they are not sited properly in the landscape.

The use of grade-change devices and drainage structures should be considered carefully in the planning and design phases of your project. With careful use they can enhance the functionality and/or aesthetic of your landscape significantly, and in many cases, help

create dramatic spaces that are useful and enjoyable throughout the year. When not considered carefully, they can be ineffective and costly in terms of resources and space.

GRADE-CHANGE DEVICES

Where limited space, existing vegetation, or other functional or aesthetic desires or conditions exist, a physical element such as a wall or step may be added to help transition from one elevation to another. The main benefit of using a grade-change device is that it allows for the greatest amount of vertical change in the shortest amount of horizontal space when compared simply to a sloping landscape.

Obviously, there are other advantages to using walls, ramps, and steps, including functional ones, such as providing seating or pedestrian access from one level to another. Aesthetic reasons also come into play; walls, for example, are among the best tools designers have for articulating a space. Still, there are some negatives to be considered, including the cost of constructing a set of stairs or a series of walls along a hillside.

WALLS—Walls allow for the greatest amount of vertical change in the least amount of horizontal distance of all grade-change devices. They are clearly more costly than a simple sloping landscape, but their sensitive and creative use can expand the usefulness of a landscape immensely.

When designing and constructing walls, consider their function in the landscape first. A freestanding wall used only to define a patio area and possibly to provide seating typically has a very different set of structural requirements than does a wall holding back a large slope or one that is creating a raised planting area.

One similarity between them all, however, is the importance of creating a solid foundation for the wall to rest upon. Regardless of a wall's form and function, its integrity depends on the adequate excavation of unstable soils and the installation of a well-drained and well-compacted subbase. This is especially true of one experiencing the kind of lateral forces that a typical retaining wall does.

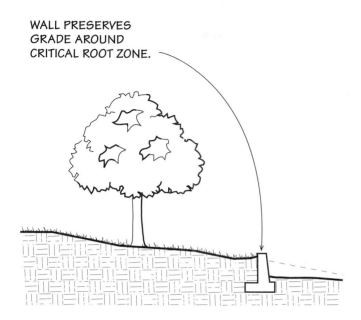

WALL PRESERVES
GRADE AROUND
CRITICAL ROOT ZONE.

PRESERVING THE GRADE AROUND TREES

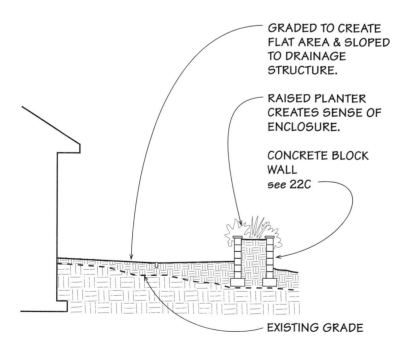

GRADED TO CREATE
FLAT AREA & SLOPED
TO DRAINAGE
STRUCTURE.

RAISED PLANTER
CREATES SENSE OF
ENCLOSURE.

CONCRETE BLOCK
WALL
see 22C

EXISTING GRADE

FREESTANDING PLANTER

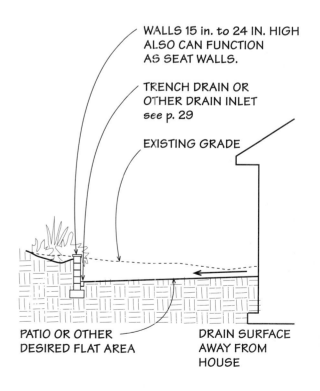

WALLS 15 in. to 24 IN. HIGH
ALSO CAN FUNCTION
AS SEAT WALLS.

TRENCH DRAIN OR
OTHER DRAIN INLET
see p. 29

EXISTING GRADE

PATIO OR OTHER
DESIRED FLAT AREA

DRAIN SURFACE
AWAY FROM
HOUSE

SLOPE RETENTION & CREATION OF
USABLE FLAT AREAS

 RETAINING WALL APPLICATIONS

The forces acting on any kind of wall must be considered carefully when designing their structure to prevent failure. These include the weight of the wall, the condition of the subsoil, the weight of the soil pressing against the back of a retaining wall, and the wind pressure built up against a freestanding wall.

Retaining walls, for example, must resist the settling of the foundation and/or subbase caused by the weight of the wall itself. It must also resist the lateral forces pressing against the back of the wall, which can cause "sliding" (the gravitational movement of the soil being retained), and rotational forces, which usually act on both the foundation and the wall itself and can cause the wall to overturn or rotate.

Drainage—Rotational failure of retaining walls is often linked to poor drainage on the uphill side of the foundation. To help avoid the level of soil-saturation that can lead to this type of failure, all retaining walls must address drainage in their design (see 21A). Foundation drains (3-in. or 4-in. perforated pipes encased in a filter fabric that are buried in gravel on the uphill side of the retaining wall) and "weep holes"

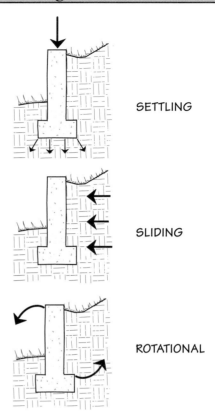

SETTLING

SLIDING

ROTATIONAL

Ⓑ   COMMON WALL FAILURES

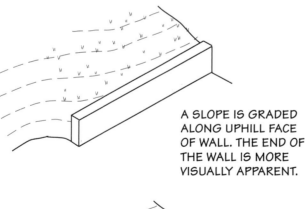

A SLOPE IS GRADED ALONG UPHILL FACE OF WALL. THE END OF THE WALL IS MORE VISUALLY APPARENT.

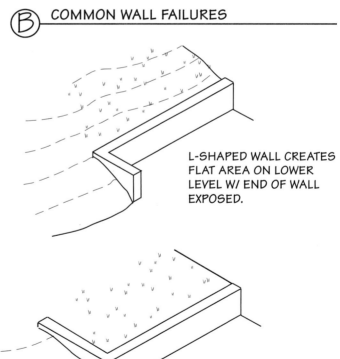

L-SHAPED WALL CREATES FLAT AREA ON LOWER LEVEL W/ END OF WALL EXPOSED.

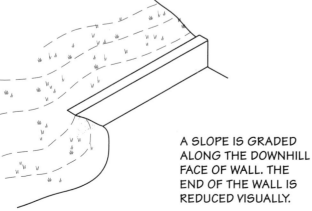

A SLOPE IS GRADED ALONG THE DOWNHILL FACE OF WALL. THE END OF THE WALL IS REDUCED VISUALLY.

L-SHAPED WALL W/ END POINTING INTO SLOPE CONCEALS END OF WALL & CREATES FLAT TERRACE ON UPPER LEVEL.

 Ⓐ   RETAINING WALLS & GRADING

(¾-in. to 1½-in. holes spaced 2 ft. to 5 ft. apart that run completely through the wall, allowing water to drain from the back of the wall to the front) are methods typically used on solid walls such as those made of concrete and masonry.

Dry-stack stone walls (walls composed of rock placed upon rock without the use of mortar), railroad tie walls, and other walls that are porous in some way typically rely on the ability of water to drain freely through the wall itself. However, walls with tight joints, such as those built with landscape timbers, are not porous enough to drain adequately despite the fact that they appear to be. These should incorporate some form of foundation drain to ensure proper drainage.

Even freestanding walls that do not retain material must still accommodate drainage. In these cases drainage is usually considered in terms of how well the subsurface foundation is drained and how the walls themselves impact surface-water runoff, i.e., whether the wall impedes drainage on the surface, causing puddling where circulation and/or other uses will take place.

Type and material selection—In addition to determining when and where to locate freestanding or retaining walls, material selection must be addressed. A wide range of options is available for all types of walls, but different materials achieve different functional and aesthetic effects. Retaining walls, for instance, require much more structural strength than freestanding walls, which can focus more narrowly on visual appeal.

Gravity walls—A type of retaining wall, gravity walls are those that by their sheer weight hold the earth in place. They frequently take the form of a large "block" of concrete poured as one mass. Another option is a dry-stack rock wall that relies on a built-in "batter," or backward vertical lean into the hillside, to help provide the force necessary to hold back the slope or soil (see 22A).

Engineered walls—Another type of retaining wall, engineered walls typically use rebar to tie the vertical "stem" of the wall structurally to the spread or cantilevered foundation. They rely on the weight of the wall and the horizontal and vertical stability of the foundation in the ground to keep the wall in place (see 22C).

Most engineered walls are made one of two ways. Concrete block can be mortared in place on a concrete foundation and reinforced with rebar that extends upward through the open cavities (or cells) in the block, which are then filled in with mortar to make a solid wall (see 22C). The other option is rebar-reinforced concrete that is "poured-in-place," meaning that it is poured into wooden forms built to suit a specific wall's circumstance (see 22D).

In either case, engineered walls are typically designed by architects, landscape architects, and engineers. They should be built with great care because the cost of dismantling and rebuilding them is much greater than repairing many other types of retaining walls, such as a dry-stack gravity wall. Any wall that will be greater than 4 ft. in height is required to be reviewed by a licensed engineer to ensure that it has been properly designed.

WEEP HOLES THAT ALLOW WATER TO PASS THROUGH ARE CRITICAL FOR SOLID WALLS.

SWALE AT TOP OF WALL MOVES WATER AWAY FROM WALL.

OPEN DRAIN ROCK

FILTER FABRIC PRESERVES ABILITY OF AGGREGATE TO SERVE DRAINAGE FUNCTION.

4-IN. PERFORATED DRAIN CARRIES WATER LATERALLY TO DRAINAGE SYSTEM.

 DRAINAGE OF RETAINING WALLS

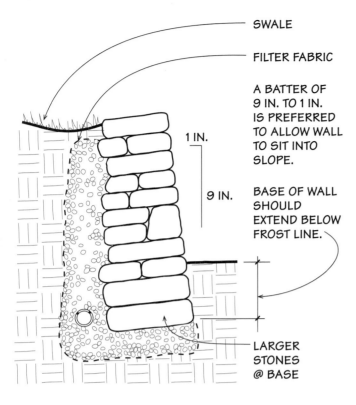

SWALE

FILTER FABRIC

A BATTER OF 9 IN. TO 1 IN. IS PREFERRED TO ALLOW WALL TO SIT INTO SLOPE.

1 IN.

9 IN.

BASE OF WALL SHOULD EXTEND BELOW FROST LINE.

LARGER STONES @ BASE

(A) DRY-STACK ROCK RETAINING WALL

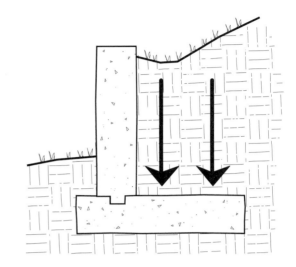

EXTENDING THE FOOTING INTO THE SLOPE ALLOWS THE FORCE OF THE SOIL TO CONTRIBUTE TO THE STRENGTH OF THE WALL.

(B) CANTILEVERED WALL FOUNDATION

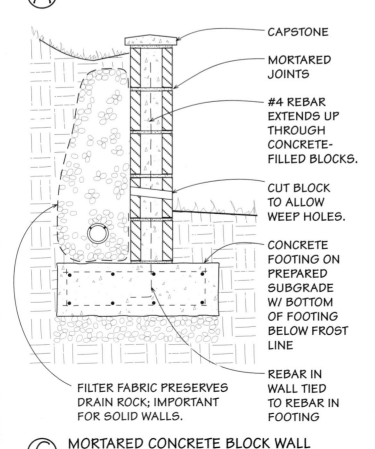

CAPSTONE

MORTARED JOINTS

#4 REBAR EXTENDS UP THROUGH CONCRETE-FILLED BLOCKS.

CUT BLOCK TO ALLOW WEEP HOLES.

CONCRETE FOOTING ON PREPARED SUBGRADE W/ BOTTOM OF FOOTING BELOW FROST LINE

REBAR IN WALL TIED TO REBAR IN FOOTING

FILTER FABRIC PRESERVES DRAIN ROCK; IMPORTANT FOR SOLID WALLS.

(C) MORTARED CONCRETE BLOCK WALL

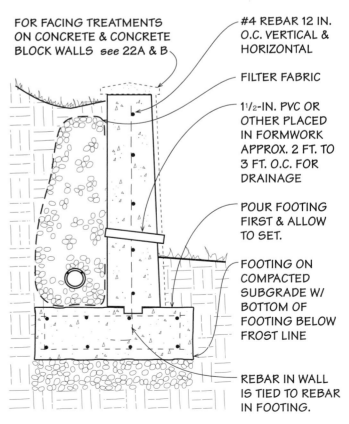

FOR FACING TREATMENTS ON CONCRETE & CONCRETE BLOCK WALLS see 22A & B

#4 REBAR 12 IN. O.C. VERTICAL & HORIZONTAL

FILTER FABRIC

1 1/2-IN. PVC OR OTHER PLACED IN FORMWORK APPROX. 2 FT. TO 3 FT. O.C. FOR DRAINAGE

POUR FOOTING FIRST & ALLOW TO SET.

FOOTING ON COMPACTED SUBGRADE W/ BOTTOM OF FOOTING BELOW FROST LINE

REBAR IN WALL IS TIED TO REBAR IN FOOTING.

(D) POURED-IN-PLACE CONCRETE WALL

Engineered walls may or may not have a batter on their face, since the bulk of their retaining ability lies within the solid foundation, the rebar embedded within the wall itself, and the weight of the soil pressing down on the base of the foundation. The face of concrete or block walls can be left exposed, or it can be covered with a veneer, usually brick, tile, rock, stucco, or faux-stone (see 23A & B). These facade treatments do little, if anything, to increase the retaining capabilities of the wall itself. They do, however, increase the aesthetic possibilities available to the homeowner and builder.

CMU retaining walls—CMUs (concrete masonry units, which are also referred to as concrete block, or sometimes, albeit inaccurately, cinder block) are becoming very prevalent in the residential landscape due to their relatively low cost, durability, and ease of installation. CMUs can be mortared or dry-stacked.

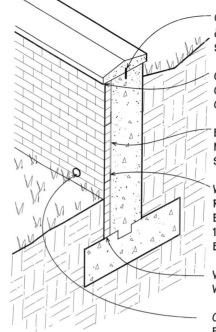

CAPSTONE MORTARED & SECURED W/ 1/4-IN. STEEL PIN.

CAPSTONE OVERHANGS ON BOTH SIDES.

BRICK VENEER MORTARED TO WALL STRUCTURE.

WIRE JOINT REINFORCEMENT EVERY 16 IN. TO 18 IN. TO SECURE BRICK TO WALL.

VENEER CARRIED TO WALL FOOTING.

CUT BRICKS TO ALLOW FOR WEEP HOLES.

NOTE:
PATTERN OF BRICK WILL VARY, DEPENDING ON AESTHETIC PREFERENCE.

NOTE:
VARIOUS ADDITIONAL WALL-FACING TREATMENTS ARE AVAILABLE. MOST ARE APPLIED W/ MORTAR & CAN BE USED ON FORMED OR BLOCK WALLS.

 BRICK VENEER

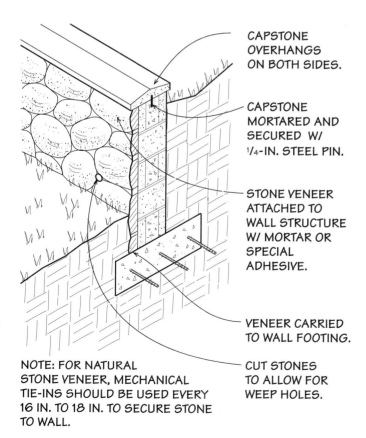

CAPSTONE OVERHANGS ON BOTH SIDES.

CAPSTONE MORTARED AND SECURED W/ 1/4-IN. STEEL PIN.

STONE VENEER ATTACHED TO WALL STRUCTURE W/ MORTAR OR SPECIAL ADHESIVE.

VENEER CARRIED TO WALL FOOTING.

CUT STONES TO ALLOW FOR WEEP HOLES.

NOTE: FOR NATURAL STONE VENEER, MECHANICAL TIE-INS SHOULD BE USED EVERY 16 IN. TO 18 IN. TO SECURE STONE TO WALL.

 MANUFACTURED STONE VENEER

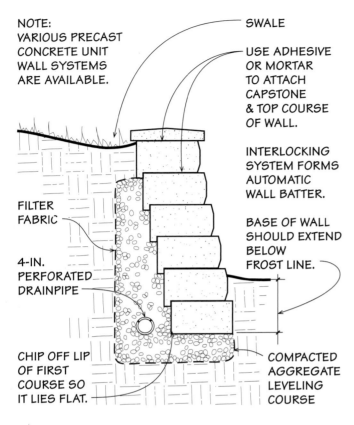

NOTE:
VARIOUS PRECAST
CONCRETE UNIT
WALL SYSTEMS
ARE AVAILABLE.

SWALE

USE ADHESIVE
OR MORTAR
TO ATTACH
CAPSTONE
& TOP COURSE
OF WALL.

INTERLOCKING
SYSTEM FORMS
AUTOMATIC
WALL BATTER.

BASE OF WALL
SHOULD EXTEND
BELOW
FROST LINE.

FILTER
FABRIC

4-IN.
PERFORATED
DRAINPIPE

CHIP OFF LIP
OF FIRST
COURSE SO
IT LIES FLAT.

COMPACTED
AGGREGATE
LEVELING
COURSE

Ⓐ INTERLOCKING CONCRETE BLOCK WALL

Dry-stock CMU walls—These walls are made with a specific type of block that is designed to lock together tightly (see 24A). Most interlocking CMU walls have a batter already built into their design.

Rock and stone walls—Rock and stone are popular materials for retaining and freestanding walls because they can be such beautiful elements in the landscape. But constructing them well is an art, requiring a lot of time, material, and skill to make the puzzle pieces of material fit together in a structurally sound yet aesthetically pleasing way.

Perhaps more than any other type, dry-stack stone walls require a marked batter in their design when used as retaining walls since it is the friction of the rocks on one another combined with the gravitational pressure the wall exerts as it leans back into the slope that keeps the wall upright and in place (see 22A). Typically, dry-stack walls above 12 in. in height should have a 2-in. to 3-in. batter for every 18 in. in height. It is important to remember to figure in the amount of horizontal distance the batter takes up, because it will potentially impact the amount of space that is available for planting or other uses at the top and/or bottom of the wall.

Timber walls—Using materials that include pressure-treated 6x6 or 6x8 railroad ties and synthetic timbers made of recycled plastic, timber walls can be very effective in shaping, embellishing, and retaining the landscape. However, wood-based material tends to decompose over time when in constant contact with the ground. Any wood material used for retaining-wall purposes, therefore, should be decay-resistant. Pressure-treated wood or the plastic- and plastic/wood-mix materials provide this kind of resistance. Even pressure-treated lumber can rot over time, however, and if ends are exposed by cutting timbers to fit, rot can be accelerated. Builders should take care to treat exposed, freshly cut ends of wooden timbers with some form of preservative during installation (see 25A).

STEPS AND RAMPS—Vertical change along a circulation route is dealt with using either steps or ramps. Stairs allow for the most rapid rise in elevation along a path of travel in the least horizontal distance, but they can be a barrier for some, such as delivery persons or people with strollers or wheelchairs.

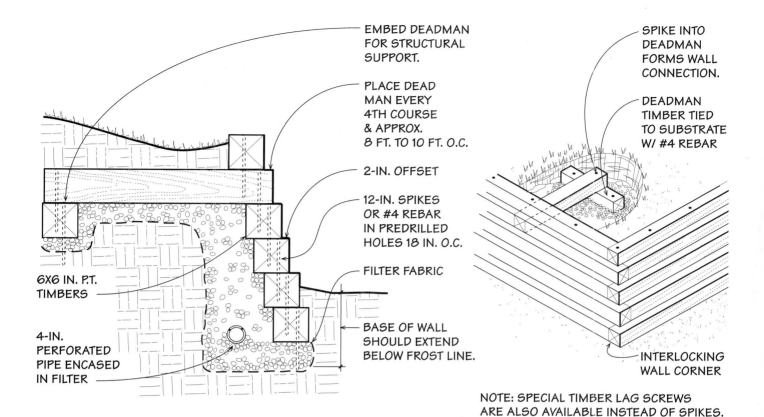

EMBED DEADMAN FOR STRUCTURAL SUPPORT.

PLACE DEAD MAN EVERY 4TH COURSE & APPROX. 8 FT. TO 10 FT. O.C.

2-IN. OFFSET

12-IN. SPIKES OR #4 REBAR IN PREDRILLED HOLES 18 IN. O.C.

FILTER FABRIC

6X6 IN. P.T. TIMBERS

4-IN. PERFORATED PIPE ENCASED IN FILTER

BASE OF WALL SHOULD EXTEND BELOW FROST LINE.

SPIKE INTO DEADMAN FORMS WALL CONNECTION.

DEADMAN TIMBER TIED TO SUBSTRATE W/ #4 REBAR

INTERLOCKING WALL CORNER

NOTE: SPECIAL TIMBER LAG SCREWS ARE ALSO AVAILABLE INSTEAD OF SPIKES.

 A HORIZONTAL TIMBER WALL

Ramps, on the other hand, favor movement by nearly everyone equally. They do, however, require a longer horizontal distance to cover the vertical change. Careful consideration of the project's context, the nature of uses in the landscape, and the home-owner's lifestyle and abilities, both now and in the future, should help identify the best solutions for a given site.

Steps—Steps may be freestanding or may be set "into" the landscape through the use of cheek walls (walls that edge the sides of the steps), which allow them to blend into their surroundings. Steps also can be built into a wall, which becomes, in effect, cheek walls as the steps move up the slope (see 26A).

Like walls, steps can be made of many different materials. Whether they are made of concrete, timbers, stone, brick, or block, all steps share basic structural characteristics. The dimensions that builders use to describe steps are referred to as the "riser" and the "tread." The riser is the vertical distance from the top surface of one step to the top surface of the next step. The tread is the horizontal distance from the front of the top surface of a step to the back of that same step. To design a flight of steps that is comfortable to walk up and down in the outdoor landscape, a simple formula is often used to determine the height of the riser and the length of the tread:

$$2R + T = 24\text{-}26 \text{ in.}$$

"R" is the height of the riser (in in.), and "T" is the length of the tread (in in.). Therefore, a step with a rise of 5 in. would require a tread length of 14 in. to 16 in. in length:

$$(2 \times 5) + 14 \text{ [or 16]} = 24 \text{ [or 26]}$$

The proportion of riser to tread can be shifted around as needed to conform to the site and its topography or the available horizontal distance in a given area. If a configuration for riser and tread that fits the formula cannot be found, then a redesign of the area or an

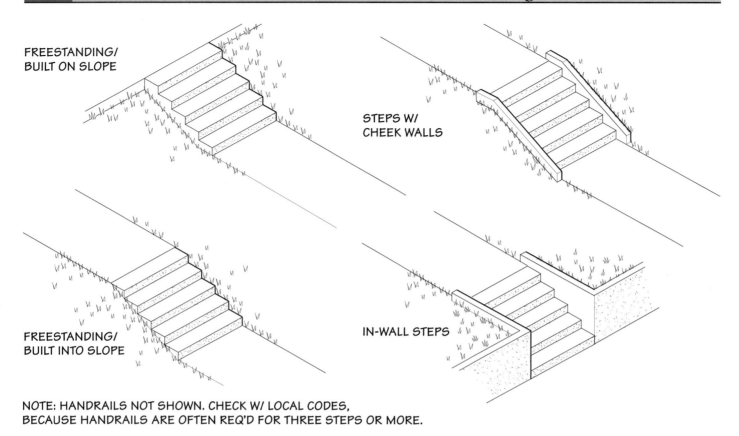

FREESTANDING/
BUILT ON SLOPE

STEPS W/
CHEEK WALLS

FREESTANDING/
BUILT INTO SLOPE

IN-WALL STEPS

NOTE: HANDRAILS NOT SHOWN. CHECK W/ LOCAL CODES,
BECAUSE HANDRAILS ARE OFTEN REQ'D FOR THREE STEPS OR MORE.

Ⓐ STEPS AND GRADING

entirely different solution may be needed. Typically, outdoor steps range from 4 in. to 7 in. in riser height, which minimizes the potential for tripping. Also critical to note is that local codes will most likely mandate that no riser may vary in height by more than ¼ in. as compared to the rest of the steps; greater variation makes for an extremely dangerous flight of steps.

Stair drainage—Drainage on and around steps is an important consideration in the design of a staircase. Typically, all steps (except wooden steps framed as part of a deck or other structure) will have a 2 percent slope from the back of the step to the front (this translates into approximately a ⅛-in. drop per lin. ft. of tread) to ensure adequate watershed. Steps should be level across the width of the tread, perpendicular to the direction of travel up and down the stairs.

Care also should be taken to ensure proper drainage at both the top and bottom of the staircase. Since puddling can lead to hazardous conditions (especially in climates where ice is an issue) water should be directed away from the bottom riser toward an open area or a drain located at least 2 ft. from the last riser,

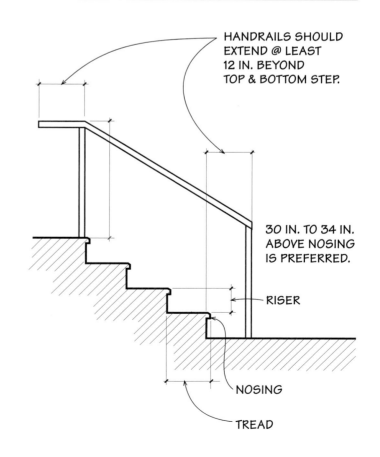

HANDRAILS SHOULD
EXTEND @ LEAST
12 IN. BEYOND
TOP & BOTTOM STEP.

30 IN. TO 34 IN.
ABOVE NOSING
IS PREFERRED.

RISER

NOSING

TREAD

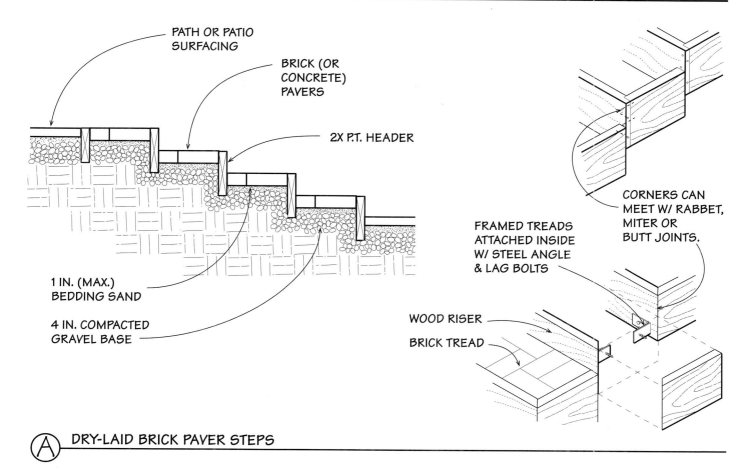

PATH OR PATIO SURFACING

BRICK (OR CONCRETE) PAVERS

2X P.T. HEADER

1 IN. (MAX.) BEDDING SAND

4 IN. COMPACTED GRAVEL BASE

FRAMED TREADS ATTACHED INSIDE W/ STEEL ANGLE & LAG BOLTS

CORNERS CAN MEET W/ RABBET, MITER OR BUTT JOINTS.

WOOD RISER

BRICK TREAD

Ⓐ DRY-LAID BRICK PAVER STEPS

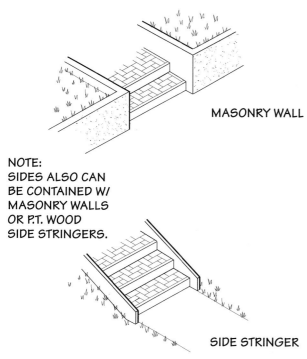

MASONRY WALL

NOTE:
SIDES ALSO CAN BE CONTAINED W/ MASONRY WALLS OR P.T. WOOD SIDE STRINGERS.

SIDE STRINGER

Ⓑ SIDE TREATMENTS FOR DRY-LAID STEPS

if at all possible. Similarly, drainage should be directed away from the top of a staircase because sheet flow could cause water to cascade down it, icing the steps.

Step traction—Stair design must take into account materials that will provide adequate traction, especially in climates prone to rain or snow. Some materials, such as concrete, can be textured easily to enhance their natural traction; other materials, such as railroad ties, can become increasingly slippery in wet weather. In some cases, a traction strip made of a nonwooden material can be added to reduce the risk of slipping, but care should be taken to evaluate the structural and aesthetic implications of adding such a material.

Handrails—Another important element to consider when designing steps is the handrail, which is required in most jurisdictions for staircases with more than three steps. The handrail should be "grippable," meaning that everyone should be able to grab it easily, and it should extend 12 in. beyond the first and last step to allow individuals the opportunity to rest or regain their balance (see 26B).

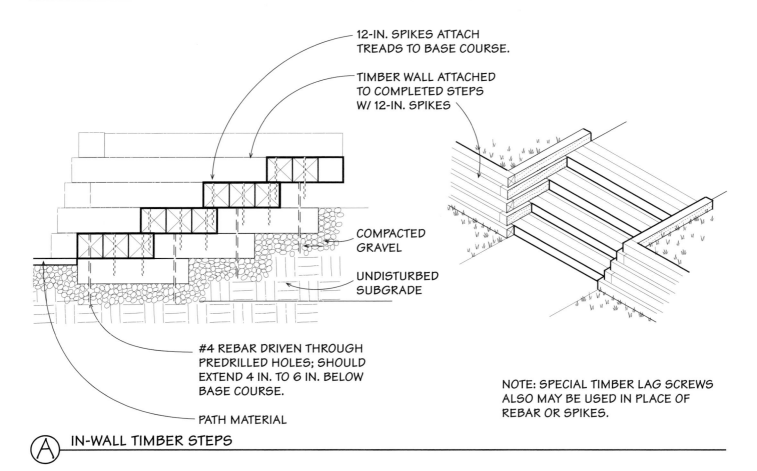

12-IN. SPIKES ATTACH TREADS TO BASE COURSE.

TIMBER WALL ATTACHED TO COMPLETED STEPS W/ 12-IN. SPIKES

COMPACTED GRAVEL

UNDISTURBED SUBGRADE

#4 REBAR DRIVEN THROUGH PREDRILLED HOLES; SHOULD EXTEND 4 IN. TO 6 IN. BELOW BASE COURSE.

PATH MATERIAL

NOTE: SPECIAL TIMBER LAG SCREWS ALSO MAY BE USED IN PLACE OF REBAR OR SPIKES.

(A) IN-WALL TIMBER STEPS

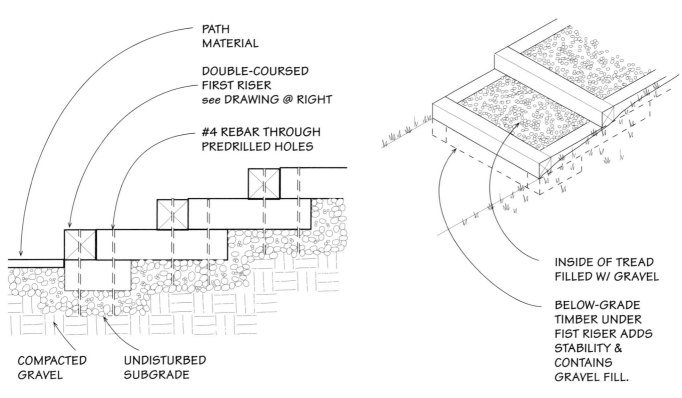

PATH MATERIAL

DOUBLE-COURSED FIRST RISER *see* DRAWING @ RIGHT

#4 REBAR THROUGH PREDRILLED HOLES

INSIDE OF TREAD FILLED W/ GRAVEL

BELOW-GRADE TIMBER UNDER FIST RISER ADDS STABILITY & CONTAINS GRAVEL FILL.

COMPACTED GRAVEL

UNDISTURBED SUBGRADE

(B) FREESTANDING TIMBER STEPS W/GRAVEL INFILL

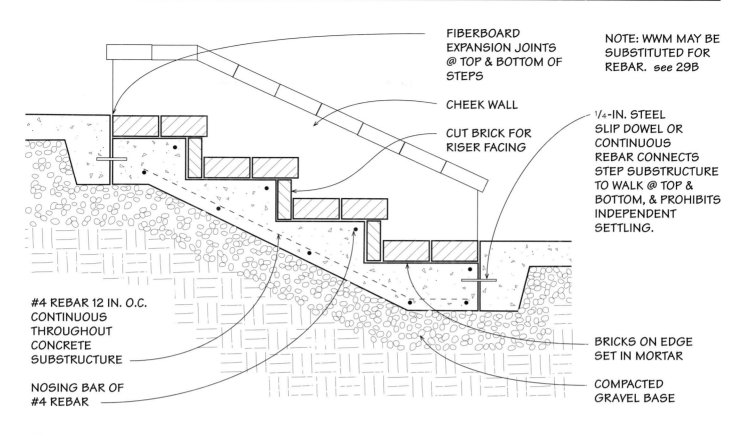

FIBERBOARD
EXPANSION JOINTS
@ TOP & BOTTOM OF
STEPS

NOTE: WWM MAY BE
SUBSTITUTED FOR
REBAR. see 29B

CHEEK WALL

CUT BRICK FOR
RISER FACING

1/4-IN. STEEL
SLIP DOWEL OR
CONTINUOUS
REBAR CONNECTS
STEP SUBSTRUCTURE
TO WALK @ TOP &
BOTTOM, & PROHIBITS
INDEPENDENT
SETTLING.

#4 REBAR 12 IN. O.C.
CONTINUOUS
THROUGHOUT
CONCRETE
SUBSTRUCTURE

NOSING BAR OF
#4 REBAR

BRICKS ON EDGE
SET IN MORTAR

COMPACTED
GRAVEL BASE

Ⓐ MORTARED BRICK STEPS W/ CHEEK WALL

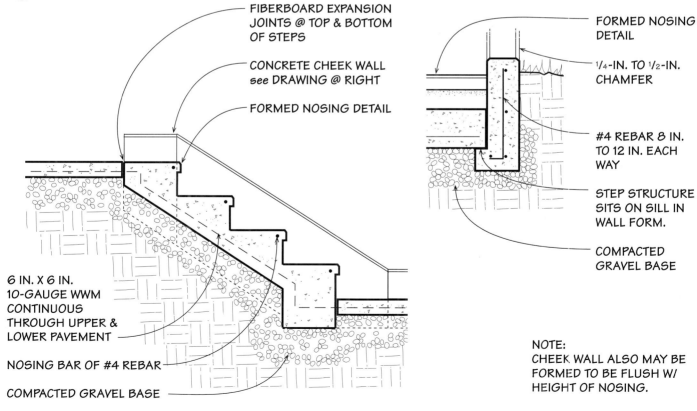

FIBERBOARD EXPANSION
JOINTS @ TOP & BOTTOM
OF STEPS

CONCRETE CHEEK WALL
see DRAWING @ RIGHT

FORMED NOSING DETAIL

FORMED NOSING
DETAIL

1/4-IN. TO 1/2-IN.
CHAMFER

#4 REBAR 8 IN.
TO 12 IN. EACH
WAY

STEP STRUCTURE
SITS ON SILL IN
WALL FORM.

COMPACTED
GRAVEL BASE

6 IN. X 6 IN.
10-GAUGE WWM
CONTINUOUS
THROUGH UPPER &
LOWER PAVEMENT

NOSING BAR OF #4 REBAR

COMPACTED GRAVEL BASE

NOTE:
CHEEK WALL ALSO MAY BE
FORMED TO BE FLUSH W/
HEIGHT OF NOSING.

Ⓑ MORTARED BRICK STEPS W/ CHEEK WALL

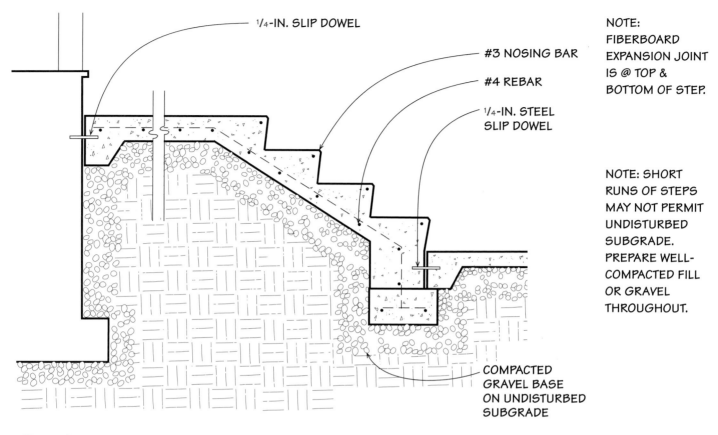

¼-IN. SLIP DOWEL

#3 NOSING BAR

#4 REBAR

¼-IN. STEEL SLIP DOWEL

NOTE: FIBERBOARD EXPANSION JOINT IS @ TOP & BOTTOM OF STEP.

NOTE: SHORT RUNS OF STEPS MAY NOT PERMIT UNDISTURBED SUBGRADE. PREPARE WELL-COMPACTED FILL OR GRAVEL THROUGHOUT.

COMPACTED GRAVEL BASE ON UNDISTURBED SUBGRADE

Ⓐ CONCRETE STEP & PORCH CONTINUOUS POUR

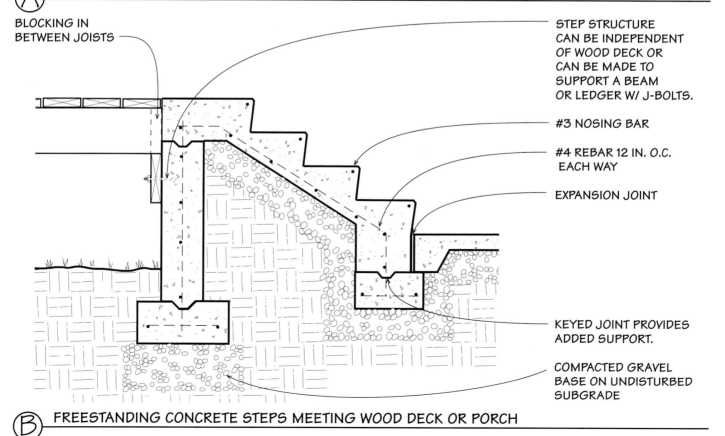

BLOCKING IN BETWEEN JOISTS

STEP STRUCTURE CAN BE INDEPENDENT OF WOOD DECK OR CAN BE MADE TO SUPPORT A BEAM OR LEDGER W/ J-BOLTS.

#3 NOSING BAR

#4 REBAR 12 IN. O.C. EACH WAY

EXPANSION JOINT

KEYED JOINT PROVIDES ADDED SUPPORT.

COMPACTED GRAVEL BASE ON UNDISTURBED SUBGRADE

Ⓑ FREESTANDING CONCRETE STEPS MEETING WOOD DECK OR PORCH

Ramps—Ramps (or sloping walkways) with a slope greater than a 1-ft. vertical rise in 20 horizontal ft. can be considered in much the same way as steps are. Both allow people to climb a slope in as comfortable and enjoyable a manner as possible and both have similar safety considerations. The obvious difference is the lack of a vertical barrier, which makes a ramp an ideal design solution that is welcoming and inclusive to all.

Ramps typically have a landing, or level area, for every 30 in. of vertical rise along the length of travel. Ramps are commonly 4 ft. wide, with 3½ ft. being the typical minimum. Their landings are generally 5 ft. to 6 ft. in length (check your local codes). Railing design is considered just as it is with step design, and drainage is similar as well, with water being pushed away from the top and bottom of the ramp whenever possible.

Unlike steps, however, a ramp can have a cross-slope, wherein the ramp itself slopes across the path of travel at a maximum of 2 percent (which is approximately ¼ in. to ½ in. for a typical ramp); this allows water to drain down the ramp but pushes it off to one side, leaving the main route of travel as puddle-free as possible (see 31A).

If considered at the outset of a design and construction project, ramps and sloping walks can be integrated into the landscape in a manner that enhances the aesthetic of the home while simultaneously increasing its livability and its openness. Incorporating planters, retaining walls, planted slopes, or even integrating ramps and landings with steps are just a few strategies you might consider when creating a landscape that will be useful, enjoyable, and accessible throughout the homeowner's life span.

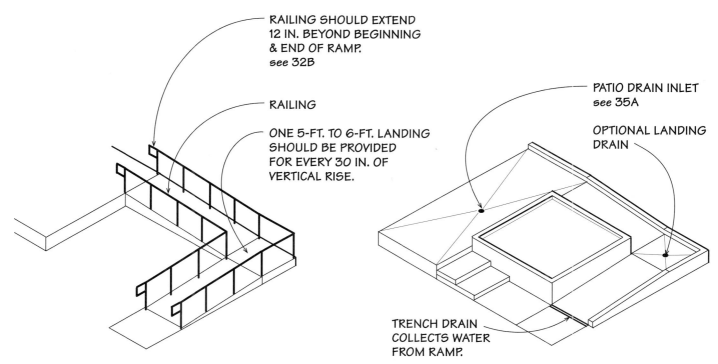

RAILING SHOULD EXTEND 12 IN. BEYOND BEGINNING & END OF RAMP. see 32B

RAILING

ONE 5-FT. TO 6-FT. LANDING SHOULD BE PROVIDED FOR EVERY 30 IN. OF VERTICAL RISE.

PATIO DRAIN INLET see 35A

OPTIONAL LANDING DRAIN

TRENCH DRAIN COLLECTS WATER FROM RAMP.

NOTE: CROSS SLOPE SHOULD BE 2% TO FACILITATE DRAINAGE, & RAMP SLOPE SHOULD NOT EXCEED 8%.

NOTE: RAMPS W/ CHEEK WALLS NEED TO ACCOUNT FOR DRAINAGE ALONG THE RUN OF THE RAMP.

 RAMP CONFIGURATIONS

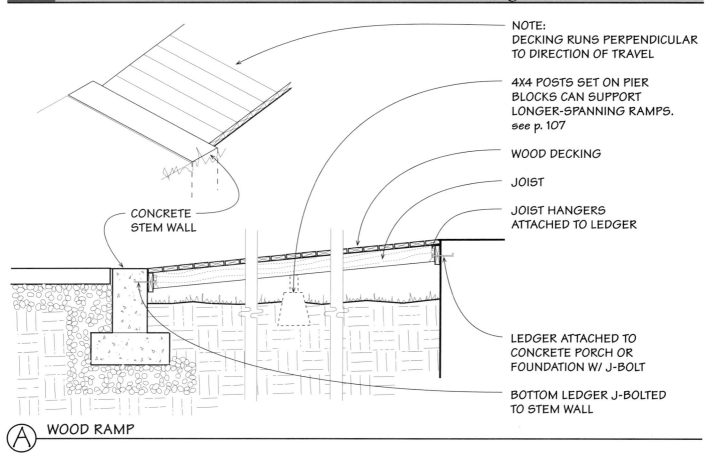

NOTE:
DECKING RUNS PERPENDICULAR
TO DIRECTION OF TRAVEL

4X4 POSTS SET ON PIER
BLOCKS CAN SUPPORT
LONGER-SPANNING RAMPS.
see p. 107

WOOD DECKING

JOIST

JOIST HANGERS
ATTACHED TO LEDGER

CONCRETE
STEM WALL

LEDGER ATTACHED TO
CONCRETE PORCH OR
FOUNDATION W/ J-BOLT

BOTTOM LEDGER J-BOLTED
TO STEM WALL

Ⓐ WOOD RAMP

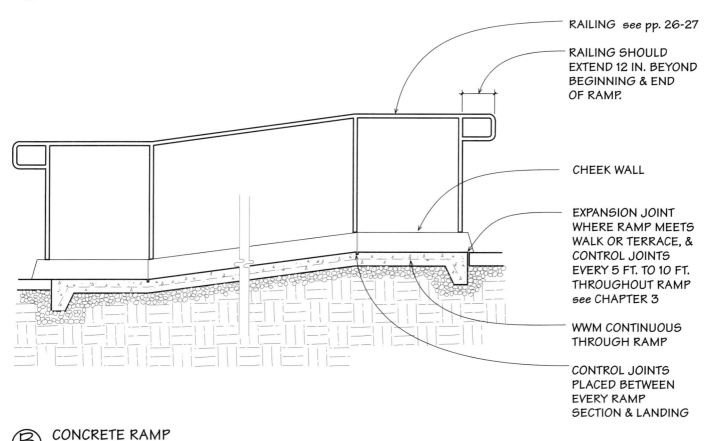

RAILING see pp. 26-27

RAILING SHOULD
EXTEND 12 IN. BEYOND
BEGINNING & END
OF RAMP.

CHEEK WALL

EXPANSION JOINT
WHERE RAMP MEETS
WALK OR TERRACE, &
CONTROL JOINTS
EVERY 5 FT. TO 10 FT.
THROUGHOUT RAMP
see CHAPTER 3

WWM CONTINUOUS
THROUGH RAMP

CONTROL JOINTS
PLACED BETWEEN
EVERY RAMP
SECTION & LANDING

Ⓑ CONCRETE RAMP

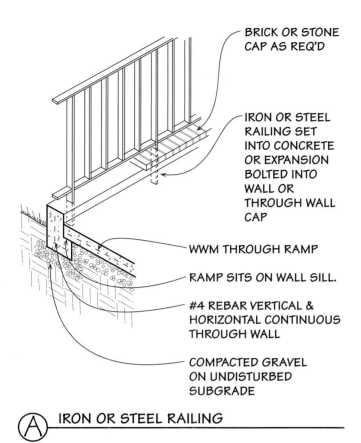

BRICK OR STONE CAP AS REQ'D

IRON OR STEEL RAILING SET INTO CONCRETE OR EXPANSION BOLTED INTO WALL OR THROUGH WALL CAP

WWM THROUGH RAMP

RAMP SITS ON WALL SILL.

#4 REBAR VERTICAL & HORIZONTAL CONTINUOUS THROUGH WALL

COMPACTED GRAVEL ON UNDISTURBED SUBGRADE

(A) IRON OR STEEL RAILING

DRAINAGE SYSTEMS

As mentioned earlier in this chapter, a primary concern in the development of a site plan is the way in which water will move over and through the site itself. Attempting to keep water out of high-use zones such as play areas, pathways, and patios requires careful consideration as you plan and construct a landscape. Fortunately, there are many options that will help you address the drainage issues on a site. However, always check with the local permitting agency to see what restrictions may exist.

BASIC CATEGORIES—There are three categories of drainage systems to consider when designing a comprehensive drainage system—open systems, closed systems, and hybrid systems, which are a combination of the first two.

Open systems—As the name implies, open systems are those that are open to the sky, such as swales, ponds, and channels. While they are typically much cheaper to construct and maintain than their closed-system counterparts, they are also more space-intensive, meaning they require outside, aboveground landscape space to work, thereby limiting other possible uses in that specific area.

Closed systems—Closed systems are typically underground and are composed of underground pipes and other elements, such as catch basins, drain inlets, trench drains, and French drains. A closed system can be a very good solution in tight landscape spaces as well as in paved areas because they allow you to maximize usable space. However, installation and maintenance costs are typically higher than open systems. Plus there can be environmental costs, including a reduction in the amount of water that percolates back into the groundwater, an increase in the speed at which runoff reaches creeks and rivers, and an increased risk of downstream flooding. For these reasons, closed systems should be considered carefully and used sparingly if possible.

Hybrid systems—Most landscapes are actually drained through a combination of open- and closed-system elements. This hybrid draws upon the strengths of each type, balancing the need to direct water away from high-use areas with the need to slow runoff and allow as much water as possible to percolate back into the soil. For example, patio areas can be drained with catch basins that convey water into a swale that runs along the side of the house, while the

Open-System Elements	Uses
Swales	Slight or more visible depression in the landscape that collects & conveys water.
Check Dam	In-swale structure designed to hold back water & allow sediments to settle out.
Ponds	Area where water collects; either can be left to percolate or can hold water w/ aid of geotextile or clay liner.
Closed-System Elements	
Drop inlet (patio drain)	Used in small areas where little debris is likely to fall.
Catch Basin	Used in larger areas, & in both landscape & patio applications.
Trench Drain	Used along bottom of steps, ramps, & areas of sheet draining.
Strip Drain (pool-deck drain)	Series of small drains used to keep sheet flow from leaving area; similar to trench drain, but smaller. Often used where paving meets steps or wall.
Hybrid Elements	
French Drain	Rock-filled trench, possibly w/ perforated pipe, used to drain landscape areas.
Sump or Dry Well	Collection point for water, subgrade hole possibly filled w/ drain rock to allow percolation.
Downspouts	Water from roof, could be collected, dispersed, or directed into pipe.

lawn areas are drained using a sheet-flow technique that also directs runoff away from the house and into the swale.

DRAINAGE-SYSTEM COMPONENTS—Depending on the specific landscape and drainage needs of a site, various elements can be used to create an effective and environmentally sound drainage system.

Drop inlets—The most simple of the closed-system components, drop inlets are typically 2-in.- to 4-in.-diameter holes in a surface that has a drain grate on top. These inlets allow water (and anything else that can fit through the holes of the grate) to flow directly into a drain pipe. Drop inlets are best used in areas where little or no debris collects on the surface being drained because they can become clogged over a relatively short period (see 35A).

Trench drains—Trench drains are prefabricated, structural, linear elements that allow sheet flow from a hard surface to be collected along a uniform line, such as at an expansion joint or at the bottom of steps or ramps (see 35B).

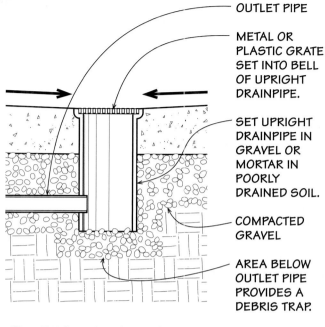

NOTE: VARIOUS INTEGRAL DROP INLETS ARE AVAILABLE. CHECK MFC.

OUTLET PIPE

METAL OR PLASTIC GRATE SET INTO BELL OF UPRIGHT DRAINPIPE.

SET UPRIGHT DRAINPIPE IN GRAVEL OR MORTAR IN POORLY DRAINED SOIL.

COMPACTED GRAVEL

AREA BELOW OUTLET PIPE PROVIDES A DEBRIS TRAP.

Ⓐ DROP INLET/PATIO DRAIN

METAL OR PLASTIC DRAIN GRATE SET INTO INTEGRAL COLLECTOR UNIT (VARIOUS TYPES ARE AVAILABLE; CHECK MFC.)

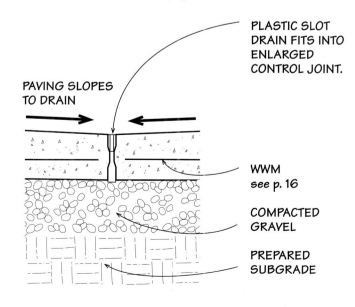

COLLECTOR UNIT TIES INTO CONCRETE PAVING.

COMPACTED GRAVEL

PREPARED SUBGRADE

NOTE: SLOT DRAINS HAVE A SMALL CAPACITY. SEVERAL DRAINS MAY BE NECESSARY. A SUBTLE WAY TO SHEET DRAIN PATIOS & POOL DECKS.

PLASTIC SLOT DRAIN FITS INTO ENLARGED CONTROL JOINT.

PAVING SLOPES TO DRAIN

WWM
see p. 16

COMPACTED GRAVEL

PREPARED SUBGRADE

Ⓑ TRENCH DRAIN

 STRIP DRAIN

NOTE: IN ADDITION TO A PLANTING BED OR LOW-LYING LAWN APPLICATION, CATCH BASINS ALSO CAN BE USED IN HARDSCAPE SITUATIONS SUCH AS PATIOS & DRIVEWAYS.

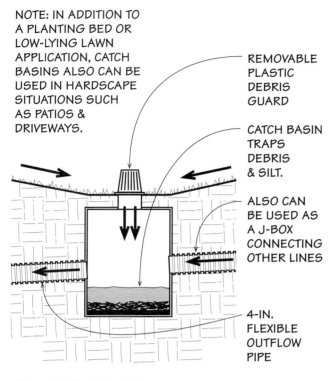

REMOVABLE PLASTIC DEBRIS GUARD

CATCH BASIN TRAPS DEBRIS & SILT.

ALSO CAN BE USED AS A J-BOX CONNECTING OTHER LINES

4-IN. FLEXIBLE OUTFLOW PIPE

 CATCH BASIN

Catch basins—Although similar to drop inlets, catch basins have a built-in sump (garbage and debris collection area beneath the outflow of the catch basin), which allows debris, such as leaf litter, to settle out in the bottom of the unit. This reduces the risk of a pipe clogging downhill of the catch basin. Periodic maintenance, including routinely cleaning out the sump, is necessary to maintain the proper function of a catch basin over time (see 36A).

French drains—A rock–filled trench, a French drain allows water to enter the cavities between the rock more easily than the surrounding soils as it directs the flow away from an area. For instance, a French drain could be used to divert water coming off a hill behind a house. French drains also can be designed to incorporate rock on the surface and perforated pipe in the subgrade gravel to help increase the amount of water being transported (see 36B).

NOTE: ALSO CAN BE BUILT W/ A 4-IN. PERFORATED PIPE @ THE BOTTOM & CAN CONNECT W/ A PIPED SYSTEM.

LAWN (ALSO CAN BE USED UNDER A VARIETY OF OTHER LANDSCAPE SITUATIONS)

FILTER FABRIC PRESERVES DRAINAGE CAPACITY OF DRAIN ROCK.

OPEN DRAIN ROCK, 1 IN. TO 3 IN. ROUND

A 4-IN. PERFORATED PIPE @ THE BOTTOM CAN ENCOURAGE DRAINAGE, ESPECIALLY IN POORLY DRAINED SOIL.

 FRENCH DRAIN

NOTE: THIS TYPE OF DRAIN SHOULD BE USED ONLY IN AREAS W/ MINIMAL DEBRIS FLOW, OR USE CATCH BASIN UNIT.

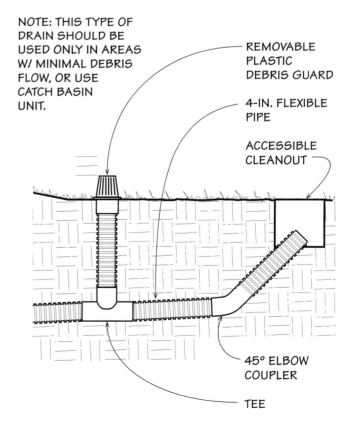

REMOVABLE PLASTIC DEBRIS GUARD

4-IN. FLEXIBLE PIPE

ACCESSIBLE CLEANOUT

45° ELBOW COUPLER

TEE

 GARDEN DRAIN ON FLEXIBLE RISER

In-ground sumps or dry wells—These allow piped water to enter a subgrade chamber (sometimes filled with cobbles or drain-rock) so that water can then percolate into the subsoil. These are best used in areas with relatively well-drained soils, although they can be of use even in poorly drained areas as a way to slow down the runoff (see 37B).

Check dams—Check dams are small structures placed within a swale that encourage water to pool up (as the name implies). This pooling allows sediments to settle out of the water while also slowing flow and encouraging percolation.

Ponds—There are two types of storm-water treatment ponds—detention ponds, which detain but do not permanently hold water back, or retention ponds, which hold water until it either percolates into the soil or evaporates into the air. Ponds can be space intensive, but their creative design can make them interesting landscape elements that are both functional and beautiful (see 38A).

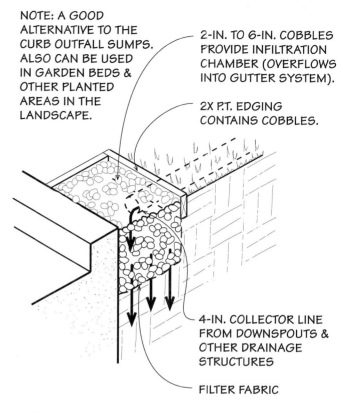

NOTE: A GOOD ALTERNATIVE TO THE CURB OUTFALL SUMPS. ALSO CAN BE USED IN GARDEN BEDS & OTHER PLANTED AREAS IN THE LANDSCAPE.

2-IN. TO 6-IN. COBBLES PROVIDE INFILTRATION CHAMBER (OVERFLOWS INTO GUTTER SYSTEM).

2X P.T. EDGING CONTAINS COBBLES.

4-IN. COLLECTOR LINE FROM DOWNSPOUTS & OTHER DRAINAGE STRUCTURES

FILTER FABRIC

Ⓑ **CURBSIDE SUMP OR DRY WELL**

4-IN. COLLECTOR LINE ALLOWS WATER TO FLOW INTO GUTTER OR OTHER IMPERVIOUS SURFACE & INTO MUNICIPAL STORM SEWER SYSTEM. CHECK CODES FOR RESTRICTIONS.

NOTE: USE STRAIGHT CURB OUTFALL ONLY WHEN SPATIALLY CONSTRAINED, IN POORLY DRAINED SOIL, OR WHEN REQUIRED BY LOCAL CODE.

Ⓐ **CURB OUTFALL**

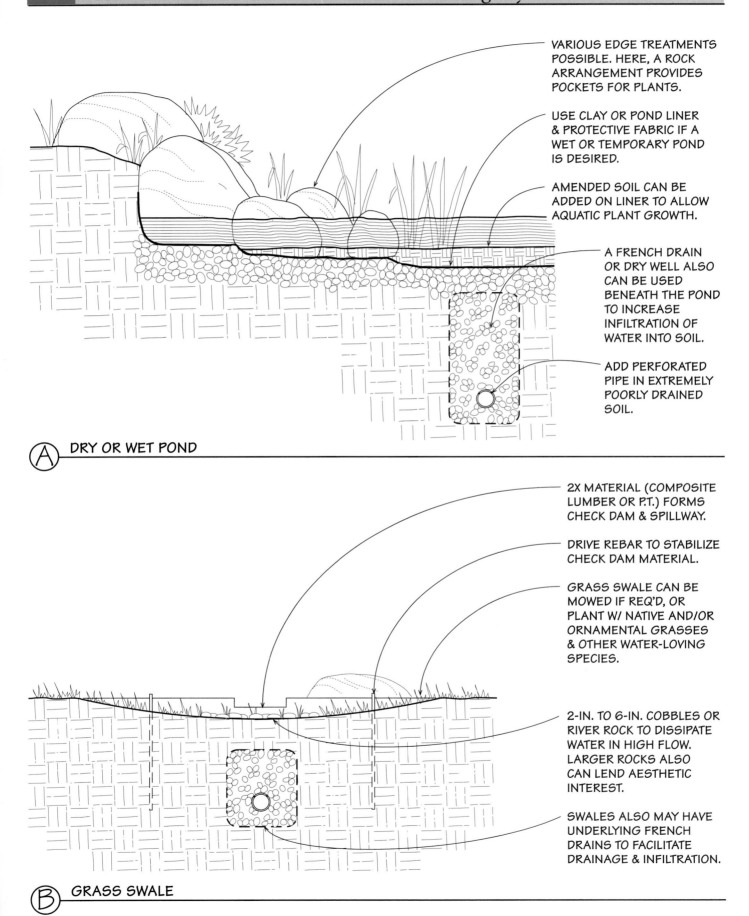

VARIOUS EDGE TREATMENTS POSSIBLE. HERE, A ROCK ARRANGEMENT PROVIDES POCKETS FOR PLANTS.

USE CLAY OR POND LINER & PROTECTIVE FABRIC IF A WET OR TEMPORARY POND IS DESIRED.

AMENDED SOIL CAN BE ADDED ON LINER TO ALLOW AQUATIC PLANT GROWTH.

A FRENCH DRAIN OR DRY WELL ALSO CAN BE USED BENEATH THE POND TO INCREASE INFILTRATION OF WATER INTO SOIL.

ADD PERFORATED PIPE IN EXTREMELY POORLY DRAINED SOIL.

(A) DRY OR WET POND

2X MATERIAL (COMPOSITE LUMBER OR P.T.) FORMS CHECK DAM & SPILLWAY.

DRIVE REBAR TO STABILIZE CHECK DAM MATERIAL.

GRASS SWALE CAN BE MOWED IF REQ'D, OR PLANT W/ NATIVE AND/OR ORNAMENTAL GRASSES & OTHER WATER-LOVING SPECIES.

2-IN. TO 6-IN. COBBLES OR RIVER ROCK TO DISSIPATE WATER IN HIGH FLOW. LARGER ROCKS ALSO CAN LEND AESTHETIC INTEREST.

SWALES ALSO MAY HAVE UNDERLYING FRENCH DRAINS TO FACILITATE DRAINAGE & INFILTRATION.

(B) GRASS SWALE

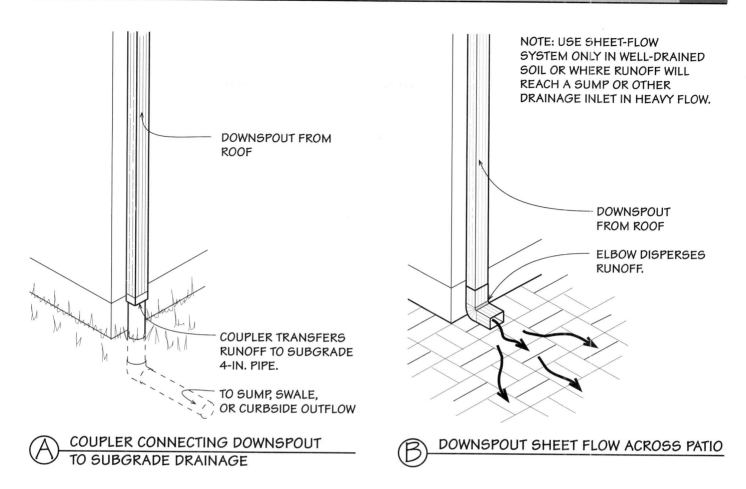

DOWNSPOUT FROM
ROOF

COUPLER TRANSFERS
RUNOFF TO SUBGRADE
4-IN. PIPE.

TO SUMP, SWALE,
OR CURBSIDE OUTFLOW

Ⓐ COUPLER CONNECTING DOWNSPOUT
TO SUBGRADE DRAINAGE

NOTE: USE SHEET-FLOW
SYSTEM ONLY IN WELL-DRAINED
SOIL OR WHERE RUNOFF WILL
REACH A SUMP OR OTHER
DRAINAGE INLET IN HEAVY FLOW.

DOWNSPOUT
FROM ROOF

ELBOW DISPERSES
RUNOFF.

Ⓑ DOWNSPOUT SHEET FLOW ACROSS PATIO

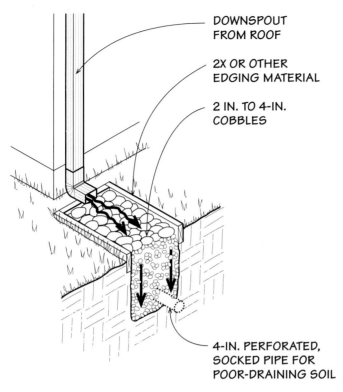

DOWNSPOUT FROM ROOF

2X OR OTHER EDGING MATERIAL

2 IN. TO 4-IN. COBBLES

4-IN. PERFORATED, SOCKED PIPE FOR POOR-DRAINING SOIL

Ⓐ DOWNSPOUT TO OPEN DRAINAGE TRENCH

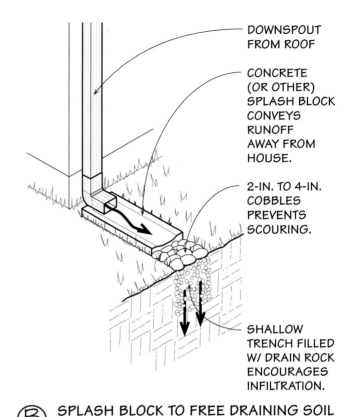

DOWNSPOUT FROM ROOF

CONCRETE (OR OTHER) SPLASH BLOCK CONVEYS RUNOFF AWAY FROM HOUSE.

2-IN. TO 4-IN. COBBLES PREVENTS SCOURING.

SHALLOW TRENCH FILLED W/ DRAIN ROCK ENCOURAGES INFILTRATION.

Ⓑ SPLASH BLOCK TO FREE DRAINING SOIL

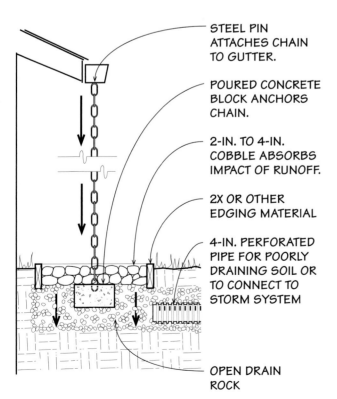

STEEL PIN ATTACHES CHAIN TO GUTTER.

POURED CONCRETE BLOCK ANCHORS CHAIN.

2-IN. TO 4-IN. COBBLE ABSORBS IMPACT OF RUNOFF.

2X OR OTHER EDGING MATERIAL

4-IN. PERFORATED PIPE FOR POORLY DRAINING SOIL OR TO CONNECT TO STORM SYSTEM

OPEN DRAIN ROCK

Ⓒ RAIN CHAIN DOWNSPOUT W/ INFILTRATION TRENCH

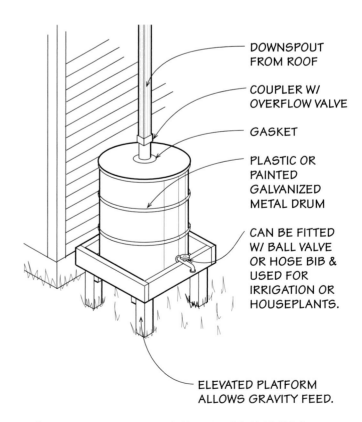

DOWNSPOUT FROM ROOF

COUPLER W/ OVERFLOW VALVE

GASKET

PLASTIC OR PAINTED GALVANIZED METAL DRUM

CAN BE FITTED W/ BALL VALVE OR HOSE BIB & USED FOR IRRIGATION OR HOUSEPLANTS.

ELEVATED PLATFORM ALLOWS GRAVITY FEED.

Ⓓ RAIN BARREL CATCHMENT SYSTEM

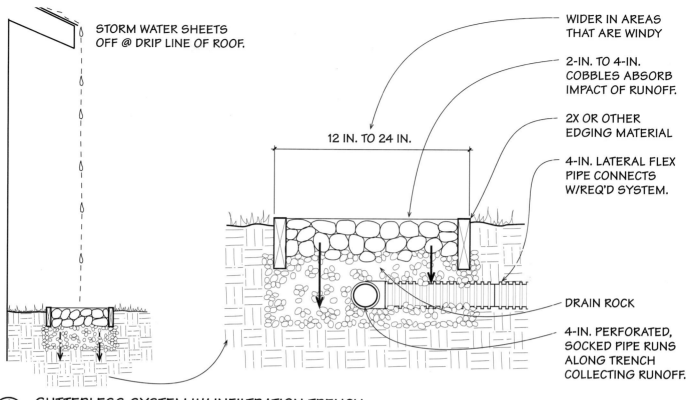

STORM WATER SHEETS
OFF @ DRIP LINE OF ROOF.

WIDER IN AREAS
THAT ARE WINDY

2-IN. TO 4-IN.
COBBLES ABSORB
IMPACT OF RUNOFF.

2X OR OTHER
EDGING MATERIAL

4-IN. LATERAL FLEX
PIPE CONNECTS
W/REQ'D SYSTEM.

12 IN. TO 24 IN.

DRAIN ROCK

4-IN. PERFORATED,
SOCKED PIPE RUNS
ALONG TRENCH
COLLECTING RUNOFF.

(A) GUTTERLESS SYSTEM W/ INFILTRATION TRENCH

CHAPTER 2

SITE UTILITIES

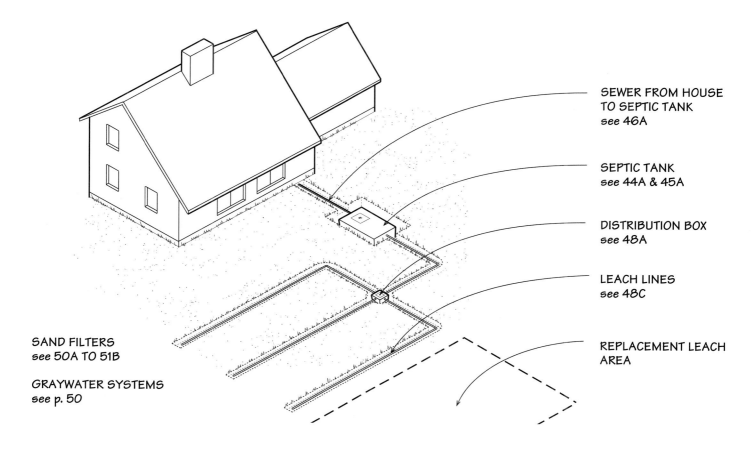

SEWER FROM HOUSE
TO SEPTIC TANK
see 46A

SEPTIC TANK
see 44A & 45A

DISTRIBUTION BOX
see 48A

LEACH LINES
see 48C

REPLACEMENT LEACH
AREA

SAND FILTERS
see 50A TO 51B

GRAYWATER SYSTEMS
see p. 50

This chapter covers the systems that are installed below ground to serve the site. Included are septic systems, irrigation systems, and lighting systems.

Septic systems tend to be a rural phenomenon and are usually installed early in the construction process, often coinciding with rough site grading. The other two systems are common in both urban and rural landscapes and are usually installed late in the construction process after finish grading and paving are complete.

SEPTIC SYSTEMS
Typical household use of water in North America averages about 65 to 75 gallons per person each day. Of this, about one-third is from toilet use, one-third is from laundry, and one-third is from sinks, tubs, and showers. Once the water has been used, it drains through waste pipes to a sewer, which exits the house to carry the wastewater (sewage) for treatment and disposal.

In urban locations, this wastewater is deposited into a municipal sewer where it flows to a sewage treatment facility. In rural locations, however, each residence has its own small sewage-treatment facility in the form of a septic system. A septic system digests most of the solid particles within the sewage and distributes the water back into the environment to recharge the groundwater supply, or aquifer.

Governmental permission to build a house often hinges on the ability of the land to support a septic system. Only after septic approval has been granted can a building permit be applied for. Septic approval is granted on a site-by-site basis and usually requires a reasonably absorbent soil in a large (approximately 1,000 sq. ft.) area that is flat or moderately sloped. The size of the house and the quality of the soil determine the size and makeup of the system.

While relatively simple in concept, septic systems are complex in practice and should generally only be designed, installed, and maintained by a licensed expert. The following will offer basic information on a system's components and requirements.

SEPTIC SYSTEM COMPONENTS–A typical residential septic system is composed of three basic parts: the septic tank, the distribution box, and the leach field. From the septic tank, the effluent flows to a distribution box that serves to distribute it evenly in several directions to a number of leach lines organized into a leach (or drain) field. In the leach field, the effluent

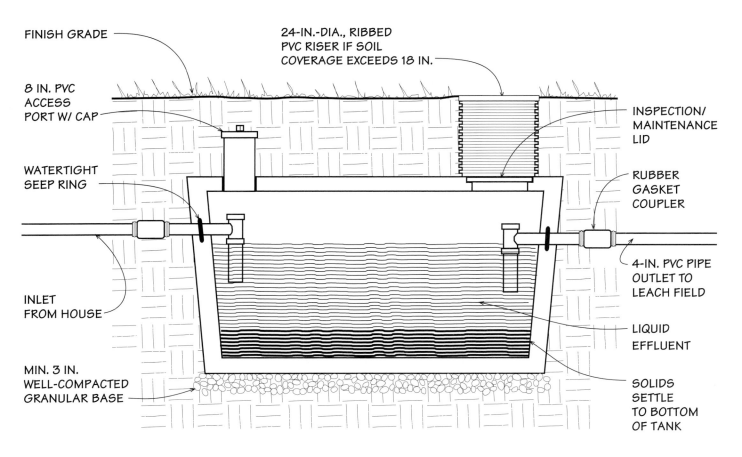

FINISH GRADE

24-IN.-DIA., RIBBED PVC RISER IF SOIL COVERAGE EXCEEDS 18 IN.

8 IN. PVC ACCESS PORT W/ CAP

INSPECTION/ MAINTENANCE LID

WATERTIGHT SEEP RING

RUBBER GASKET COUPLER

INLET FROM HOUSE

4-IN. PVC PIPE OUTLET TO LEACH FIELD

LIQUID EFFLUENT

MIN. 3 IN. WELL-COMPACTED GRANULAR BASE

SOLIDS SETTLE TO BOTTOM OF TANK

Ⓐ STANDARD CONCRETE SEPTIC TANK

flows into gravel beds through perforated lines and either evaporates into the atmosphere or recharges the groundwater by percolating into the soil.

The septic tank—The septic tank is a large (1,000- to 1,500-gallon), buried chamber—usually made of concrete or fiberglass—that receives the sewage waste directly as it flows out of the house (see 44A & 45A). Sewage waste is composed mostly of water plus some solid waste in the form of fecal material, toilet paper, and kitchen scraps. In the septic tank, anaerobic bacteria decompose the solid waste, converting most of it to liquid, which, when mixed with the water in the sewage, is called effluent.

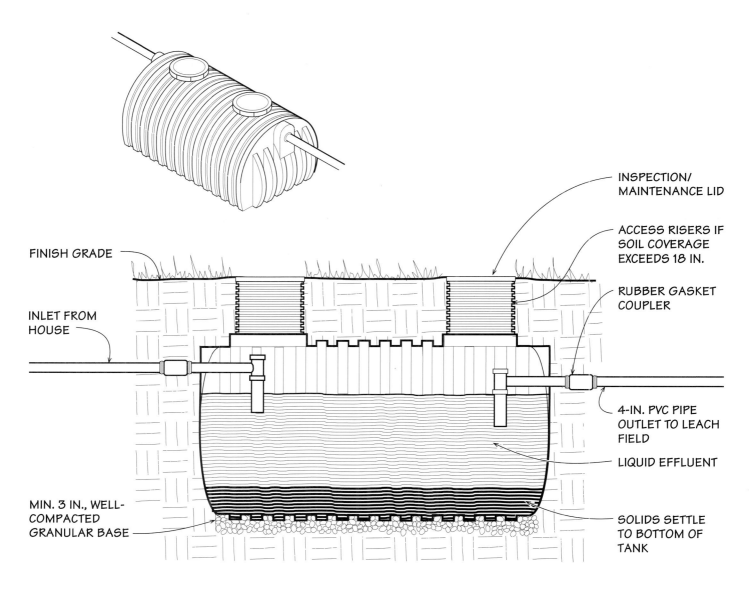

INSPECTION/ MAINTENANCE LID

ACCESS RISERS IF SOIL COVERAGE EXCEEDS 18 IN.

RUBBER GASKET COUPLER

FINISH GRADE

INLET FROM HOUSE

4-IN. PVC PIPE OUTLET TO LEACH FIELD

LIQUID EFFLUENT

MIN. 3 IN., WELL-COMPACTED GRANULAR BASE

SOLIDS SETTLE TO BOTTOM OF TANK

Ⓐ STANDARD PLASTIC SEPTIC TANK

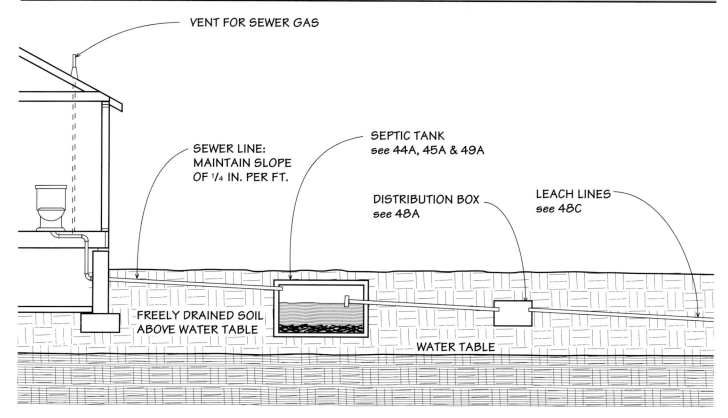

VENT FOR SEWER GAS

SEWER LINE: MAINTAIN SLOPE OF 1/4 IN. PER FT.

SEPTIC TANK see 44A, 45A & 49A

DISTRIBUTION BOX see 48A

LEACH LINES see 48C

FREELY DRAINED SOIL ABOVE WATER TABLE

WATER TABLE

 SEWAGE DISPOSAL SYSTEM

The sewer pipe that connects the house to the septic tank must slope at least 1/4 in. per ft., so the inlet to a septic tank that is 100 ft. from the house will have to be 25 in. below the level of the sewer pipe as it passes through the house foundation (see 46A).

Distribution box and drain field—The distribution box and drain field work together to disperse the effluent over a large area, where it is recycled back into the environment (see 47A & 47B). The ideal location for a drain field is downhill from the house, away from streams and wells, and not under a driveway or other road. A drain field is typically required to be set back a minimum of 50 ft. to 100 ft. from a lake or stream, 100 ft. from a well, and 20 ft. from a dwelling.

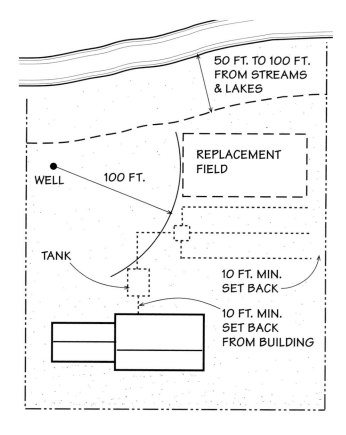

50 FT. TO 100 FT. FROM STREAMS & LAKES

REPLACEMENT FIELD

WELL

100 FT.

TANK

10 FT. MIN. SET BACK

10 FT. MIN. SET BACK FROM BUILDING

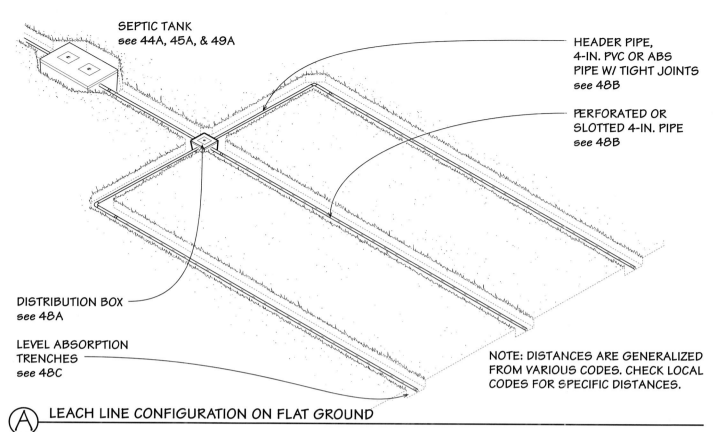

SEPTIC TANK
see 44A, 45A, & 49A

HEADER PIPE,
4-IN. PVC OR ABS
PIPE W/ TIGHT JOINTS
see 48B

PERFORATED OR
SLOTTED 4-IN. PIPE
see 48B

DISTRIBUTION BOX
see 48A

LEVEL ABSORPTION
TRENCHES
see 48C

NOTE: DISTANCES ARE GENERALIZED
FROM VARIOUS CODES. CHECK LOCAL
CODES FOR SPECIFIC DISTANCES.

A LEACH LINE CONFIGURATION ON FLAT GROUND

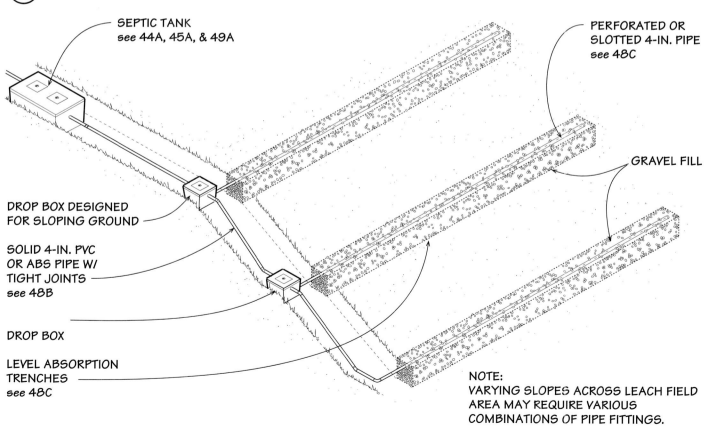

SEPTIC TANK
see 44A, 45A, & 49A

PERFORATED OR
SLOTTED 4-IN. PIPE
see 48C

GRAVEL FILL

DROP BOX DESIGNED
FOR SLOPING GROUND

SOLID 4-IN. PVC
OR ABS PIPE W/
TIGHT JOINTS
see 48B

DROP BOX

LEVEL ABSORPTION
TRENCHES
see 48C

NOTE:
VARYING SLOPES ACROSS LEACH FIELD
AREA MAY REQUIRE VARIOUS
COMBINATIONS OF PIPE FITTINGS.

 B LEACH LINE CONFIGURATION ON SLOPING GROUND

NOTE:
DISTRIBUTION BOXES ARE AVAILABLE
IN PLASTIC OR CONCRETE.

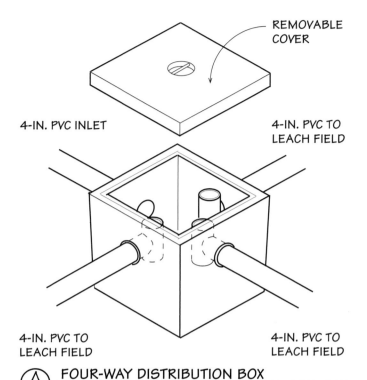

REMOVABLE COVER

4-IN. PVC INLET

4-IN. PVC TO LEACH FIELD

4-IN. PVC TO LEACH FIELD

4-IN. PVC TO LEACH FIELD

Ⓐ FOUR-WAY DISTRIBUTION BOX

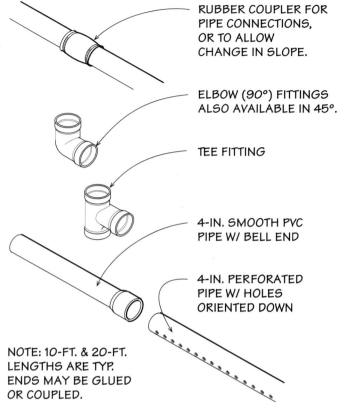

RUBBER COUPLER FOR PIPE CONNECTIONS, OR TO ALLOW CHANGE IN SLOPE.

ELBOW (90°) FITTINGS ALSO AVAILABLE IN 45°.

TEE FITTING

4-IN. SMOOTH PVC PIPE W/ BELL END

4-IN. PERFORATED PIPE W/ HOLES ORIENTED DOWN

NOTE: 10-FT. & 20-FT. LENGTHS ARE TYP. ENDS MAY BE GLUED OR COUPLED.

Ⓑ PVC OR ABS PIPE COMPONENTS

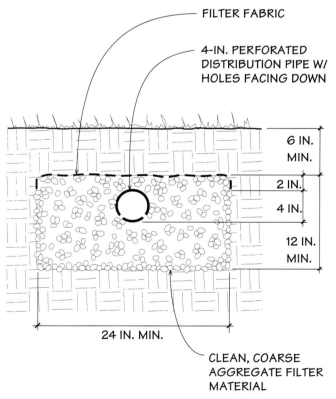

FILTER FABRIC

4-IN. PERFORATED DISTRIBUTION PIPE W/ HOLES FACING DOWN

6 IN. MIN.

2 IN.

4 IN.

12 IN. MIN.

24 IN. MIN.

CLEAN, COARSE AGGREGATE FILTER MATERIAL

Ⓒ LEACH LINE IN ABSORPTION TRENCH

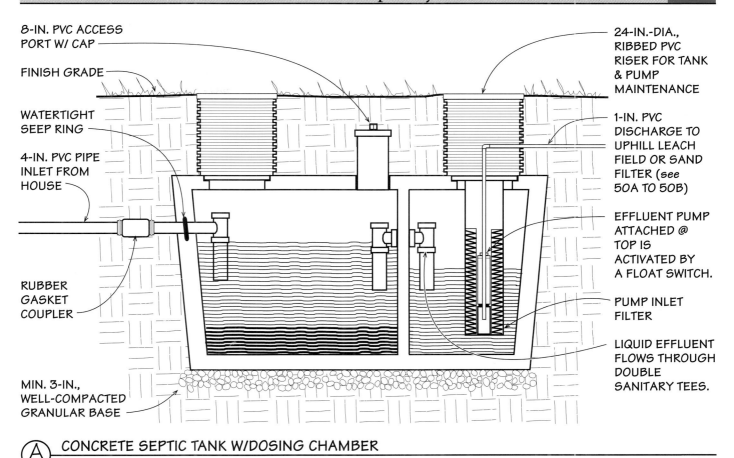

8-IN. PVC ACCESS PORT W/ CAP

FINISH GRADE

WATERTIGHT SEEP RING

4-IN. PVC PIPE INLET FROM HOUSE

RUBBER GASKET COUPLER

MIN. 3-IN., WELL-COMPACTED GRANULAR BASE

24-IN.-DIA., RIBBED PVC RISER FOR TANK & PUMP MAINTENANCE

1-IN. PVC DISCHARGE TO UPHILL LEACH FIELD OR SAND FILTER (see 50A TO 50B)

EFFLUENT PUMP ATTACHED @ TOP IS ACTIVATED BY A FLOAT SWITCH.

PUMP INLET FILTER

LIQUID EFFLUENT FLOWS THROUGH DOUBLE SANITARY TEES.

Ⓐ CONCRETE SEPTIC TANK W/DOSING CHAMBER

Dosing tank—When the only location for a drain field is uphill from the house, the effluent can be pumped from the septic tank to the distribution box. This is accomplished by using a two-chamber septic tank, called a dosing tank; digestion occurs in the first chamber, and the second chamber contains only the liquid effluent with a pump to force it uphill (see 49A).

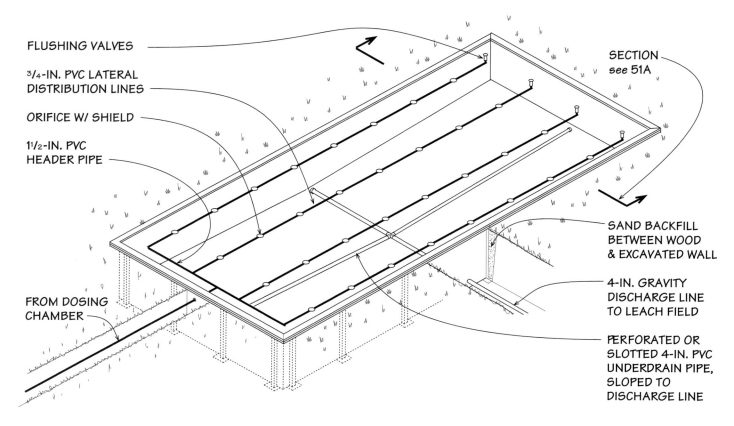

FLUSHING VALVES

¾-IN. PVC LATERAL DISTRIBUTION LINES

ORIFICE W/ SHIELD

1½-IN. PVC HEADER PIPE

FROM DOSING CHAMBER

SECTION see 51A

SAND BACKFILL BETWEEN WOOD & EXCAVATED WALL

4-IN. GRAVITY DISCHARGE LINE TO LEACH FIELD

PERFORATED OR SLOTTED 4-IN. PVC UNDERDRAIN PIPE, SLOPED TO DISCHARGE LINE

 SAND FILTER

Sand filter system—When the soil is not sufficiently porous to allow the percolation of effluent from a traditional drain field, a sand filter system may be employed (see 50A, 51A & 51B). With this system, a thick layer of sand is used to purify the effluent prior to its distribution in the drain field.

The effluent is pumped through small-diameter pipes to the top of the sand bed, where it is forced through orifices and descends to the bottom of the bed. As it descends, the effluent is filtered by the sand, then collected at the base of the sand bed by 4-in. perforated pipes that route it to a leach field. Although a leach field and replacement field (in case the original field fails) are still required, a sand filter can reduce the necessary size of the field by up to 50 percent.

Both traditional and sand filter drain fields must be kept free of heavy traffic (such as farm equipment or livestock) which would compact the soil, and must be free of vegetation, which has roots that could potentially disrupt the drain lines.

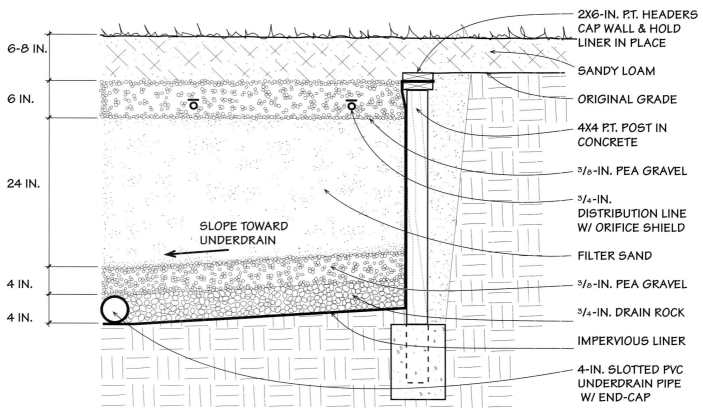

6-8 IN.

6 IN.

24 IN.

SLOPE TOWARD
UNDERDRAIN

4 IN.

4 IN.

2X6-IN. P.T. HEADERS
CAP WALL & HOLD
LINER IN PLACE

SANDY LOAM

ORIGINAL GRADE

4X4 P.T. POST IN
CONCRETE

³/₈-IN. PEA GRAVEL

³/₄-IN.
DISTRIBUTION LINE
W/ ORIFICE SHIELD

FILTER SAND

³/₈-IN. PEA GRAVEL

³/₄-IN. DRAIN ROCK

IMPERVIOUS LINER

4-IN. SLOTTED PVC
UNDERDRAIN PIPE
W/ END-CAP

(A) SAND FILTER
SECTION

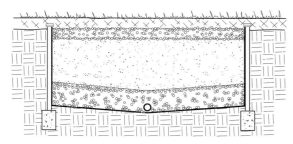

FULLY BURIED W/ FULL CONTAINMENT

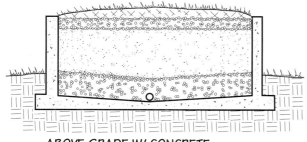

ABOVE GRADE W/ CONCRETE
CONTAINMENT STRUCTURE

NOTE: THESE OPTIONS REPRESENT THE EXTREMES.
OTHER PARTIALLY BURIED OPTIONS ALSO EXIST.

(B) SAND FILTER CONFIGRUATION OPTIONS

GRAYWATER SYSTEMS—An alternative to the standard septic system is a graywater system in which toilet waste is separated from other plumbing wastes, and the treatment process deals only with more benign wastewater from sinks, showers, and laundry facilities. The treated water can then be used on the property to irrigate gardens, recharge ponds, or flush toilets. Human waste must, of course, be dealt with in other acceptable ways, such as with a composting toilet.

Compared to septic systems, the greatest advantage of graywater systems is that their use allows houses to be built on sites with soil unsuitable for a leach field. In addition, they're more environmentally responsible because they recycle water. Graywater systems are more expensive and complicated than septic systems, however, and they require more maintenance.

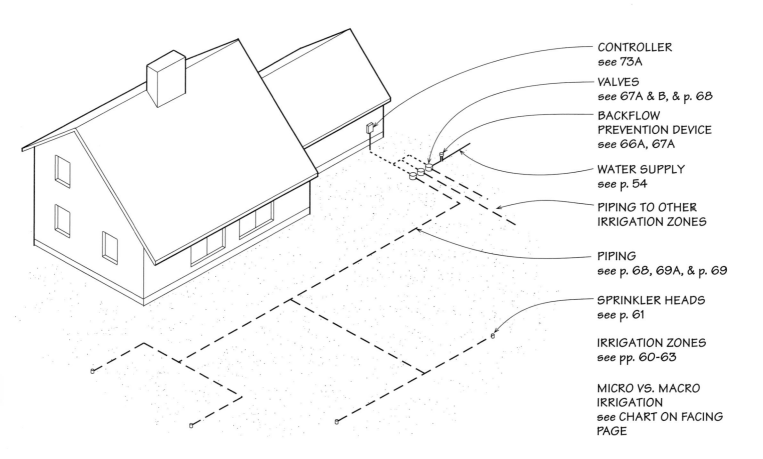

CONTROLLER
see 73A

VALVES
see 67A & B, & p. 68

BACKFLOW
PREVENTION DEVICE
see 66A, 67A

WATER SUPPLY
see p. 54

PIPING TO OTHER
IRRIGATION ZONES

PIPING
see p. 68, 69A, & p. 69

SPRINKLER HEADS
see p. 61

IRRIGATION ZONES
see pp. 60-63

MICRO VS. MACRO
IRRIGATION
see CHART ON FACING
PAGE

LANDSCAPE IRRIGATION SYSTEMS

A significant advantage of irrigation systems is that they eliminate the need for hand watering; an underground sprinkler system can water an entire yard automatically. The convenience of this type of system is coupled with the aesthetic benefit of its underground location. There are no hoses, and sprinkler heads can retract, becoming almost invisible when not in use.

Sprinkler systems are also water-efficient. An automatic controller can schedule the system to operate at times when water loss due to evaporation and wind is least likely to occur. The controller also eliminates over- and underwatering due to human forgetfulness. Irrigation systems are typically organized into zones that contain plants with common watering needs, conserving water and creating a healthier garden.

The requirements for a modern irrigation system are elementary:

• a fresh water supply (well or municipal) with adequate pressure and flow rate. (It is best if the water is free of sediment and harsh chemicals, but these can be removed with filters);

• electrical power to operate a programmable controller;

• permits when required in many jurisdictions.

A new irrigation system is most conveniently installed after all grading has been done and large plants such as trees and large shrubs are in the ground. At this point in construction, it is relatively easy with a trenching machine to dig the narrow trenches required for underground piping, and the sprinkler heads can be set in the ground at their finished height.

For an existing yard, installation of an underground irrigation system is also relatively simple because modern trenching equipment causes minimal damage to existing lawns or beds.

Since the piping for an irrigation system consists of simple plastic pipe connections that can be glued or clamped together, much of the work can be executed by the homeowner.

THE BASIC IRRIGATION SYSTEM—Residential irrigation systems are conveniently divided into two basic types—macro irrigation and micro irrigation.

Macro irrigation is the more traditional system, and it's the type commonly used to irrigate lawns and other large zones in residential gardens using sprinkler heads or flood bubblers.

Micro irrigation (also called drip irrigation) is a more recent development, which uses lower pressure and smaller lines than macro systems to irrigate small areas close to the ground in a more precise way. It is especially effective with plants that require irrigation only during establishment. In arid regions, entire projects are irrigated with micro irrigation because of its higher water efficiency.

	MACRO IRRIGATION	**MICRO IRRIGATION**
Typical Use	Lawns and medium to large planting beds	Small or odd-shaped lawns and beds, narrow and linear beds, and pots
Water Efficiency	65%–80%	80%–100%
Cost	Initial cost similar to micro: materials slightly higher; installation labor slightly lower.	Initial cost similar to macro: materials slightly lower; installation labor slightly higher. Long-term maintenance higher than macro.
Maintenance	Little	Clogging can be a problem due to small orifices and low pressure.
Aesthetics	Piping below grade and pop-up heads can retract to be nearly invisible.	Heads and drip lines can be above ground, but small and not very noticeable.
Throw Distance	Bubbler or 8 ft.-30 ft.+	Spot (drip) or 2 ft.-10 ft.
Pressure and Flow	High pressure and high flow	Low pressure and low flow
Pipe Size	1½ in.-½ in.	½ in.-⅛ in.
Examples	Water guns, impact-driven heads, multiple stream rotors, spray pop-ups, flood bubblers	Bubblers, drip emitters, in-line drips, soaker lines, low-volume sprayers

Both macro and micro irrigation begin with the same basic system. Pressurized water is piped in a large-diameter pipe buried below the frost line. This pipe is connected to a collection of valves that control water flow to the various zones in the system. Each of the valves, which are opened and closed electrically via a programmable control panel, regulates water flow to one zone. Piping extends from the valves to each of the zones, where sprinkler heads and other watering devices distribute the water to the garden.

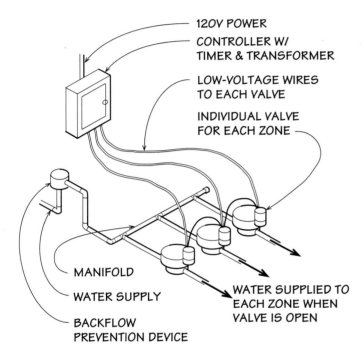

120V POWER

CONTROLLER W/ TIMER & TRANSFORMER

LOW-VOLTAGE WIRES TO EACH VALVE

INDIVIDUAL VALVE FOR EACH ZONE

MANIFOLD

WATER SUPPLY

BACKFLOW PREVENTION DEVICE

WATER SUPPLIED TO EACH ZONE WHEN VALVE IS OPEN

A backflow prevention device prohibits the contamination of domestic water by water from the irrigation system. Without a backflow prevention device, contamination could occur when the municipal water supply suddenly develops negative pressure, due to pumping by the fire department, for example. Most jurisdictions require a spring-loaded vacuum breaker or double check valve as a backflow prevention device. The device must be installed on the pressure side of the zone valves (see 55A).

In cold climates, automatic drains empty the irrigation lines of water after each use so that freezing weather does not damage them.

Water supply—Not surprisingly, a proper water supply is critical for an effective irrigation system. Most residential irrigation systems are able to operate using the same municipal water supply that feeds the plumbing inside the house, and water from a well or cistern also can be pumped for use. Regardless of the source, however, the water must have adequate pressure, flow rate, and quality.

Pressure—Municipal water typically is delivered with a pressure ranging from 25 to 50 pounds per square inch (psi). The pressure required for irrigation is approximately 35 psi for macro, 20 psi for micro. Inadequate pressure can be increased with the addition of a pump, and excessive pressure can be reduced with a pressure regulator.

Flow rate—Municipal water generally has a flow rate adequate to supply indoor and outdoor needs simultaneously. Most irrigation systems require a flow rate of 10 gallons per minute (gpm) or more. For systems that draw from the same water supply as the house, a much greater flow rate must be available or else the irrigation system, when operating, will disrupt water flow inside the house. Likewise, water use in the house can affect the performance of the irrigation system if the flow rate is inadequate. With a very low flow rate, a micro system is often the only practical answer.

Water quality—Especially for well and cistern systems, the water supply should be checked for dissolved chemicals that can build up sediments in pipes and fittings and for suspended particulates that can clog lines. Both of these concerns are particularly problematic in micro irrigation systems but can be controlled with the addition of filters to the system.

Valve system—At the head of an irrigation system is a manifold with a collection of valves that control the flow of water to each of the zones (see 55A & B). The valves operate on low-voltage electricity regulated by the controller. Typically, only one valve is open at a time, allowing water to flow to one zone in the system.

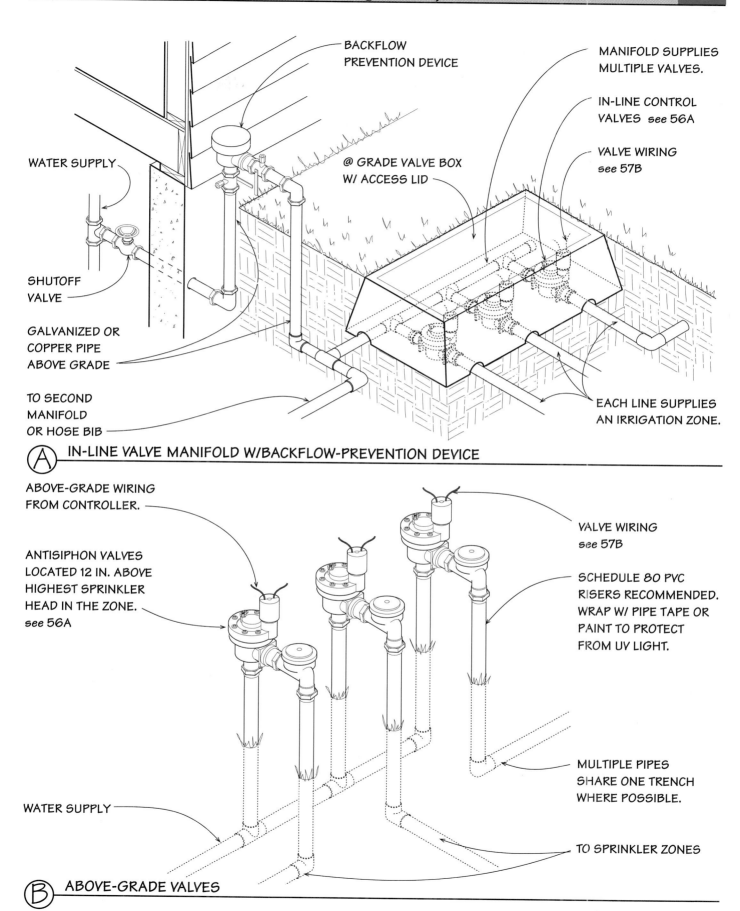

BACKFLOW
PREVENTION DEVICE

MANIFOLD SUPPLIES
MULTIPLE VALVES.

IN-LINE CONTROL
VALVES see 56A

VALVE WIRING
see 57B

WATER SUPPLY

@ GRADE VALVE BOX
W/ ACCESS LID

SHUTOFF
VALVE

GALVANIZED OR
COPPER PIPE
ABOVE GRADE

TO SECOND
MANIFOLD
OR HOSE BIB

EACH LINE SUPPLIES
AN IRRIGATION ZONE.

Ⓐ IN-LINE VALVE MANIFOLD W/BACKFLOW-PREVENTION DEVICE

ABOVE-GRADE WIRING
FROM CONTROLLER.

ANTISIPHON VALVES
LOCATED 12 IN. ABOVE
HIGHEST SPRINKLER
HEAD IN THE ZONE.
see 56A

VALVE WIRING
see 57B

SCHEDULE 80 PVC
RISERS RECOMMENDED.
WRAP W/ PIPE TAPE OR
PAINT TO PROTECT
FROM UV LIGHT.

MULTIPLE PIPES
SHARE ONE TRENCH
WHERE POSSIBLE.

WATER SUPPLY

TO SPRINKLER ZONES

Ⓑ ABOVE-GRADE VALVES

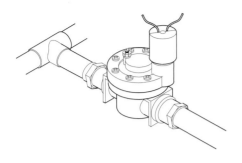

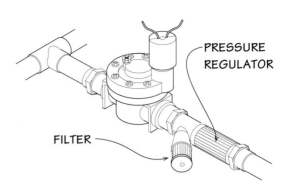

PRESSURE
REGULATOR

FILTER

IN-LINE VALVE

TYPICALLY INSTALLED ALONG
A MANIFOLD BELOW GRADE.
MUST BE PRECEDED BY
BACKFLOW PREVENTION
DEVICE. ACCESS TO
VALVES IS PROVIDED BY
PLASTIC VALVE BOX.
see 67A

ANTISIPHON VALVE

ANTISIPHON VALVES ACT AS
BACKFLOW PREVENTION
DEVICES ON THEIR OWN.
THEREFORE, A SEPARATE
BACKFLOW PREVENTION
DEVICE IS NOT NEEDED.
VALVES SHOULD BE AT LEAST
12 IN. ABOVE THE HIGHEST
SPRINKLER HEAD IN THE ZONE.
THUS, THEY NORMALLY
ARE INSTALLED ABOVE GROUND.
see 67B

IN-LINE VALVE FOR
MICRO-IRRIGATION ZONE

MICRO-IRRIGATION SYSTEMS
DEMAND A LOWER PRESSURE,
& PARTICLES CAN CLOG THE
SMALL OPENINGS OF THE HEADS.
THUS, IN-LINE VALVES FITTED W/
FILTERS & PRESSURE REGULATORS
ARE OFTEN USED. VALVES FOR
MICRO-IRRIGATION ZONES CAN
SHARE THE SAME MANIFOLD
W/ MACRO-IRRIGATION ZONES.

 VALVE TYPES

Manifolds—The manifold should be located centrally in the garden and as close as possible to the water source and controller. A single central location will minimize the amount of pipe and simplify the routing of control wires. Alternatively, a garden may have several manifolds with a smaller number of valves at each one. A pressurized water supply line (accompanied by control wires) must link the manifolds, but this allows hose bibs to be conveniently located in the garden.

Manifolds are unsightly, so most are positioned below grade in a protective box. The below-grade location also protects piping and wiring from the elements and makes insulation against freezing relatively effortless. Manifolds are sometimes located above grade (in a hidden location) because this affords easier access for the initial construction, remodeling, and occasional maintenance (see 55B).

Controllers—Controllers regulate the watering cycles for each of the irrigation zones by sending low-voltage electrical signals to the control valves. Typically, the controller opens only one valve at any one time so that no two zones are ever activated simultaneously. Controllers are either electromechanical or digital and most can be programmed for seven- or 14-day cycles. Features such as ground moisture sensors or rain shutoffs can be integrated with some controllers to make irrigation more efficient (see 57A & B).

When selecting a controller, it is wise to choose one that allows multiple programs and extra terminals to accommodate future expansion of the irrigation system. Controllers that provide multiple programs allow greater flexibility in delivering the appropriate water to specific areas.

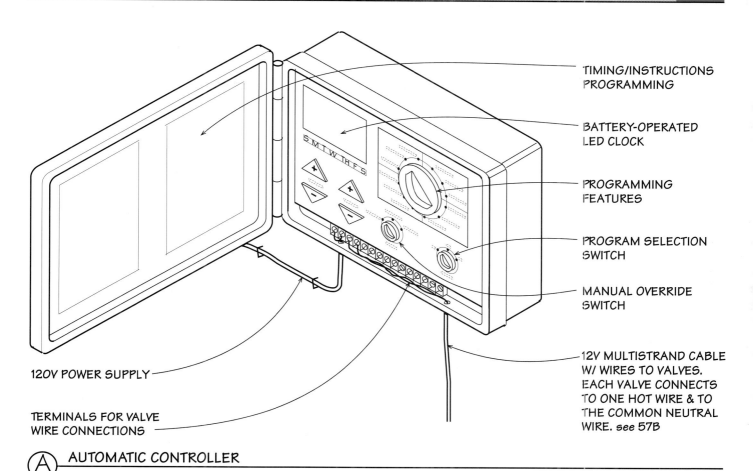

TIMING/INSTRUCTIONS
PROGRAMMING

BATTERY-OPERATED
LED CLOCK

PROGRAMMING
FEATURES

PROGRAM SELECTION
SWITCH

MANUAL OVERRIDE
SWITCH

12V MULTISTRAND CABLE
W/ WIRES TO VALVES.
EACH VALVE CONNECTS
TO ONE HOT WIRE & TO
THE COMMON NEUTRAL
WIRE. see 57B

120V POWER SUPPLY

TERMINALS FOR VALVE
WIRE CONNECTIONS

(A) AUTOMATIC CONTROLLER

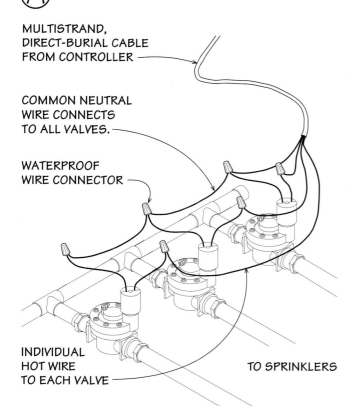

MULTISTRAND,
DIRECT-BURIAL CABLE
FROM CONTROLLER

COMMON NEUTRAL
WIRE CONNECTS
TO ALL VALVES.

WATERPROOF
WIRE CONNECTOR

INDIVIDUAL
HOT WIRE
TO EACH VALVE

TO SPRINKLERS

(B) VALVE WIRING

Piping to the zones—The piping between valves and watering zones typically consists of 1-in., ¾-in., or ½-in. rigid polyvinyl chloride (PVC) or flexible polyethylene (poly) pipe buried 8 in. to 12 in. below grade. PVC is more common because it is easier to work with. Pipe diameter must also be considered because it affects water pressure.

PVC pipe—PVC pipe is available in a grade called schedule 40, which is commonly used for house plumbing, or a thinner-walled grade called class 200. The schedule 40 pipe costs about 40 percent more than class 200. It is typical to use schedule 40 pipe for portions of the system that are under constant pressure and class 200 pipe for lateral lines in the zones. The strongest PVC pipe is schedule 80, which is sometimes used above grade for risers to elevated sprinkler heads (see 58A).

Poly pipe—Poly pipe is available in psi ratings that range from 80 psi to 160 psi. For above-grade installations, poly pipe has an advantage over PVC because it is not degraded by ultraviolet light, although its use is limited to zone lines (not main lines) (see 58B).

Piping and water pressure—Because larger diameter pipe has much greater flow capacity than smaller pipe (for example, ¾-in. pipe has approximately two times the flow capacity of ½-in. pipe), it is used for larger zones or zones that are more remote from the zone valves.

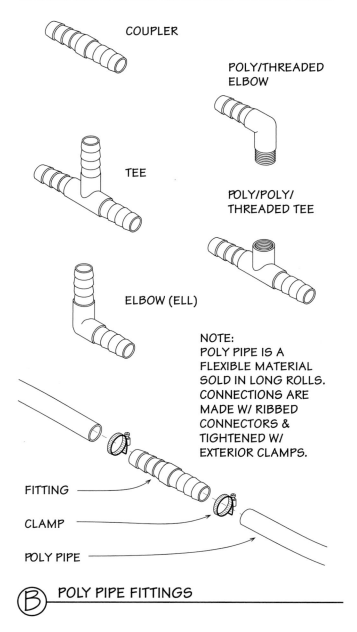

COUPLER

POLY/THREADED ELBOW

TEE

POLY/POLY/ THREADED TEE

ELBOW (ELL)

NOTE:
POLY PIPE IS A FLEXIBLE MATERIAL SOLD IN LONG ROLLS. CONNECTIONS ARE MADE W/ RIBBED CONNECTORS & TIGHTENED W/ EXTERIOR CLAMPS.

FITTING

CLAMP

POLY PIPE

SLIP/SLIP 90 (ELL OR ELBOW)

THREADED/ SLIP 90 (ELL OR ELBOW)

SLIP COUPLER

SLIP/ SLIP/ THREADED TEE

SLIP/ SLIP/ SLIP TEE

THREADED/ SLIP COUPLER

SLIP/SLIP REDUCING BUSHING

THREADED RISER

AUTOMATIC DRAIN VALVE THREADED

MALE & FEMALE LENGTHS OF PIPE

NOTE: RIGID PVC PIPE IS NORMALLY SCHEDULE 40 & COMES IN 10-FT. OR 20-FT. LENGTHS. CONNECTIONS ARE TYP. MADE W/ PVC CEMENT.

Ⓐ PVC PIPE FITTINGS

Ⓑ POLY PIPE FITTINGS

Piping should be designed to provide approximately equal pressure to the heads in each zone. Water pressure decreases along the length of a pipe due to friction loss, so one useful strategy is to branch the pipes from manifold to head in such a way that each pipe is roughly the same length (see 59A). Fittings can decrease pressure as well, so the number of fittings should be factored when considering the balanced flow of water to heads within a zone. For irrigation on hillsides, zones are usually organized horizontally along the contours to avoid the pressure differential that would occur between heads at different elevations.

ELBOW OR TEE

ALL PIPING IN ZONE IS SLOPED TO DRAIN TO THIS LOWEST POINT.

VALVE ALLOWS WATER TO DRAIN ONLY WHEN THERE IS NO PRESSURE IN THE PIPE.

GRAVEL ALLOWS PIPE TO DRAIN.

NOTE: DRAIN VALVES CAN BE @ END OF LINES OR IN THE MIDDLE BUT SHOULD ALWAYS BE @ THE LOWEST POINT IN THE ZONE.

Ⓑ AUTOMATIC DRAIN VALVES

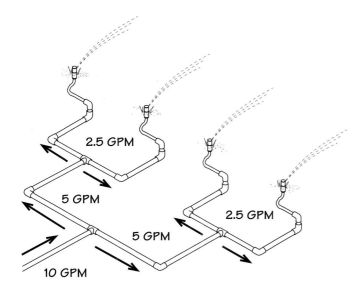

2.5 GPM

5 GPM

2.5 GPM

5 GPM

10 GPM

NOTE: DIVIDING WATER FLOW ALLOWS EVEN DISTRIBUTION BY SPRINKLER HEADS.

Ⓐ BRANCHING FOR LATERAL PIPES

IRRIGATION ZONES—Gardens are typically composed of many kinds of plants that have different watering needs. Lawns, shrubs, flower beds, trees, and vegetable gardens each require different amounts of water and thrive with different watering methods. Lawns, for example, are most easily watered from above with a general spray, while rose bushes do not tolerate water on their leaves and are best watered by soaking the ground.

Frequency of watering is another variable that must be considered. Annuals may require watering every day or two for a short period, while trees will benefit from long and infrequent watering. Gardens also have microclimates and varying soil conditions, all of which require distinct watering patterns.

To accomplish the different amounts and kinds of watering required in a garden, irrigation systems are divided into zones of plants with similar watering needs. Each zone contains several sprinkler heads or other watering devices that can be operated simultaneously with the available water pressure.

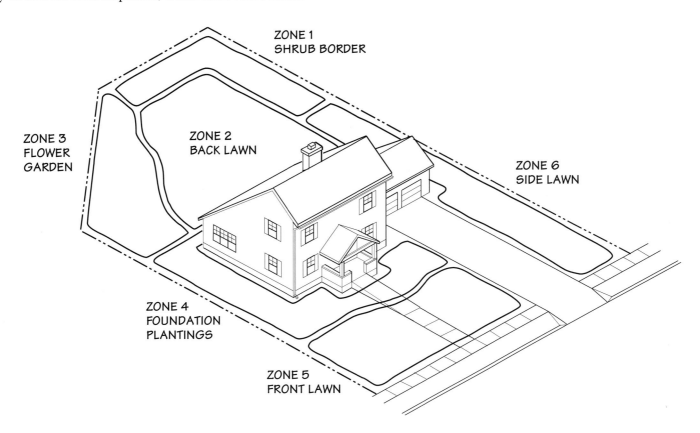

ZONE 1
SHRUB BORDER

ZONE 3
FLOWER
GARDEN

ZONE 2
BACK LAWN

ZONE 6
SIDE LAWN

ZONE 4
FOUNDATION
PLANTINGS

ZONE 5
FRONT LAWN

Designing the Zones—Once the zones are organized based on plant types, microclimates, and contours, it must be determined whether macro or micro irrigation will be used in each zone. It is common to have both macro and micro zones in the same garden and fed from different valves on the same manifold, but the two types of systems should not be mixed within the same zone because they invariably operate with different water pressure and apply water at different rates.

Spray heads are generally spaced so that the spray overlaps in order to obtain complete coverage of an area. The angle of coverage can be controlled at the spray head to direct the water at the plants rather than on pathways or buildings adjacent to the garden.

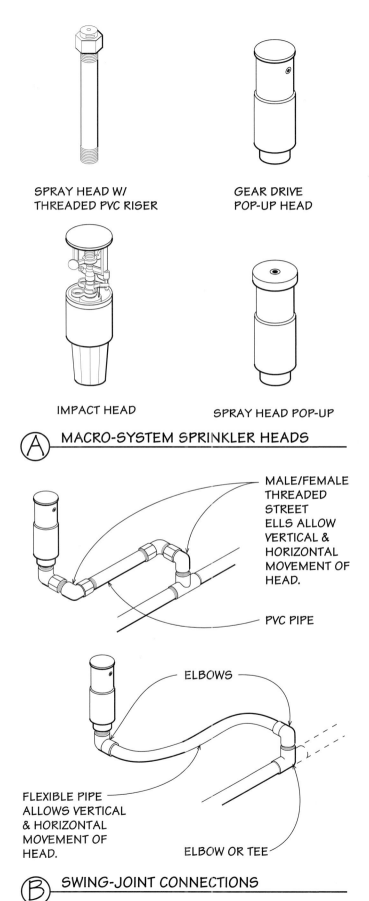

SPRAY HEAD W/ THREADED PVC RISER

GEAR DRIVE POP-UP HEAD

IMPACT HEAD

SPRAY HEAD POP-UP

(A) MACRO-SYSTEM SPRINKLER HEADS

MALE/FEMALE THREADED STREET ELLS ALLOW VERTICAL & HORIZONTAL MOVEMENT OF HEAD.

PVC PIPE

ELBOWS

FLEXIBLE PIPE ALLOWS VERTICAL & HORIZONTAL MOVEMENT OF HEAD.

ELBOW OR TEE

(B) SWING-JOINT CONNECTIONS

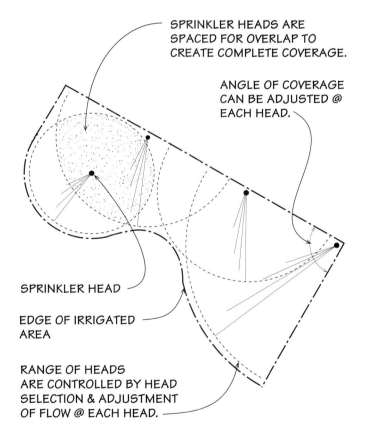

SPRINKLER HEADS ARE SPACED FOR OVERLAP TO CREATE COMPLETE COVERAGE.

ANGLE OF COVERAGE CAN BE ADJUSTED @ EACH HEAD.

SPRINKLER HEAD

EDGE OF IRRIGATED AREA

RANGE OF HEADS ARE CONTROLLED BY HEAD SELECTION & ADJUSTMENT OF FLOW @ EACH HEAD.

Macro systems—Macro systems are used extensively to irrigate lawns, planting beds, and other large garden areas. Residential-scale spray heads typically have a range of 15 ft. or more, so they can be spaced at considerable distance from one another. Heads should be located so that the spray of one head reaches the adjacent head for full coverage (see 61A & B).

Micro systems—Micro systems are used for efficiency or for long rows of plants or narrow beds that are difficult to irrigate with macro systems. Micro systems are designed to distribute water precisely at the roots of plants. Underground emitter systems have less evaporation than any other system.

Within each zone, the supply piping for micro-irrigation systems (in the form of flexible poly tubing) is typically located on the surface of the ground below a layer of mulch. A regulator can be incorporated into the system if water pressure is too great.

There is a wide range of devices to distribute the water in a micro system, including soaker hoses, drip emitters, micro bubblers, and micro sprayers (see 62A, 63A & B).

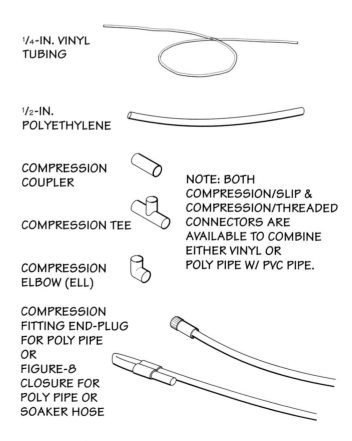

¼-IN. VINYL TUBING

½-IN. POLYETHYLENE

COMPRESSION COUPLER

COMPRESSION TEE

NOTE: BOTH COMPRESSION/SLIP & COMPRESSION/THREADED CONNECTORS ARE AVAILABLE TO COMBINE EITHER VINYL OR POLY PIPE W/ PVC PIPE.

COMPRESSION ELBOW (ELL)

COMPRESSION FITTING END-PLUG FOR POLY PIPE OR FIGURE-8 CLOSURE FOR POLY PIPE OR SOAKER HOSE

 MICRO-IRRIGATION COMPONENTS

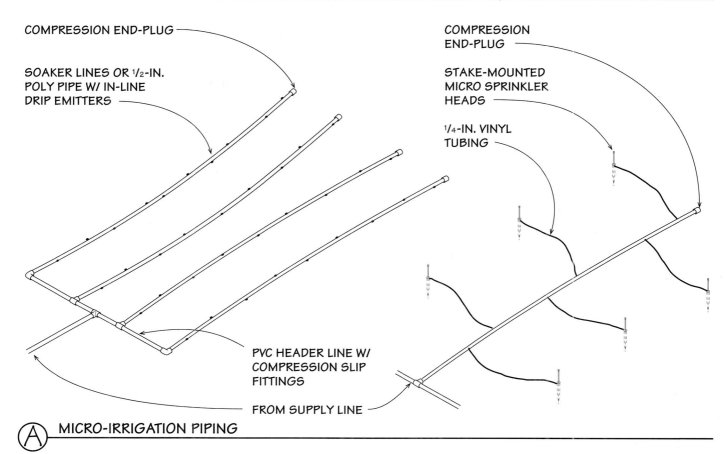

COMPRESSION END-PLUG

SOAKER LINES OR 1/2-IN. POLY PIPE W/ IN-LINE DRIP EMITTERS

COMPRESSION END-PLUG

STAKE-MOUNTED MICRO SPRINKLER HEADS

1/4-IN. VINYL TUBING

PVC HEADER LINE W/ COMPRESSION SLIP FITTINGS

FROM SUPPLY LINE

Ⓐ MICRO-IRRIGATION PIPING

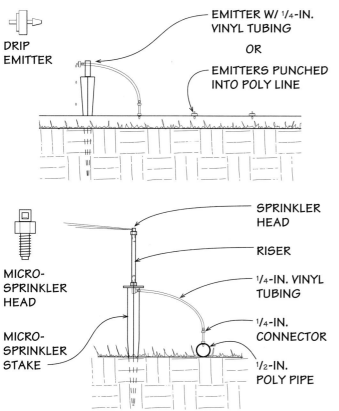

DRIP EMITTER

EMITTER W/ 1/4-IN. VINYL TUBING

OR

EMITTERS PUNCHED INTO POLY LINE

SPRINKLER HEAD

RISER

1/4-IN. VINYL TUBING

1/4-IN. CONNECTOR

1/2-IN. POLY PIPE

MICRO-SPRINKLER HEAD

MICRO-SPRINKLER STAKE

 MICRO-IRRIGATION SPRINKLERS

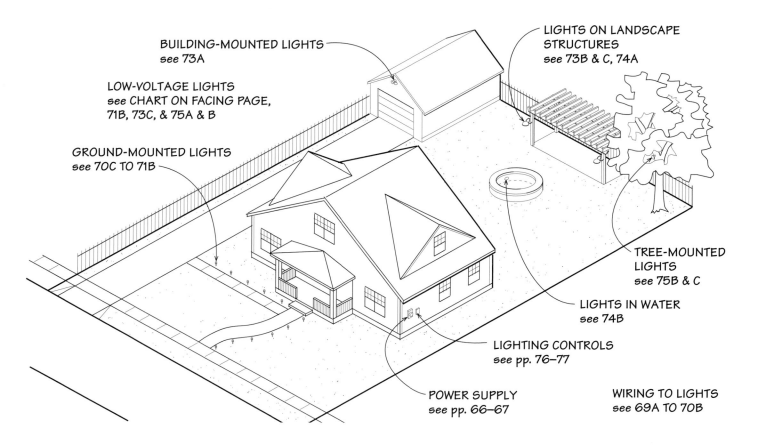

BUILDING-MOUNTED LIGHTS
see 73A

LOW-VOLTAGE LIGHTS
see CHART ON FACING PAGE,
71B, 73C, & 75A & B

GROUND-MOUNTED LIGHTS
see 70C TO 71B

LIGHTS ON LANDSCAPE
STRUCTURES
see 73B & C, 74A

TREE-MOUNTED
LIGHTS
see 75B & C

LIGHTS IN WATER
see 74B

LIGHTING CONTROLS
see pp. 76–77

POWER SUPPLY
see pp. 66–67

WIRING TO LIGHTS
see 69A TO 70B

LANDSCAPE LIGHTING

Landscape lighting can contribute to the safety, usefulness, and beauty of a site. Path lighting allows people to see where they are going at night, and security lights can be a deterrent to intruders. Lighting at decks, terraces, and lawns allows social events to extend into the nighttime hours. Lighting of recreational courts and fields permits their use under the stars. When lighting is done for aesthetic affect, such as showcasing a specimen tree, it can transform a garden at night, complementing its daytime splendor.

	ADVANTAGES	DISADVANTAGES
LINE-VOLTAGE	• Switching is simpler, so greater control is possible. • More light is available for security or nighttime activities. • Lower voltage drop • Can be easily integrated with other line-voltage site requirements, such as remote receptacles, fans, gate openers, etc.	• The voltage is high enough to be very dangerous, so great care must be taken to promote safety during and after installation. • The wiring must be buried very deep and placed in conduit when above ground. This adds significantly to installation cost. • Energy use is higher than for low-voltage systems. • Each fixture must be connected with a junction box, which is an eyesore if it can't be buried.
LOW-VOLTAGE	• Low-voltage wiring can be near or at the surface of the ground, so installation is much less expensive than for line-voltage systems. • Energy use is much lower than for line-voltage systems.	• A transformer is required to step the (line) voltage down to 12 volts. • Switching/control of low-voltage lines is less flexible than 120-volt systems. • Low-voltage fixtures tend to be less durable than line-voltage fixtures.

OUTDOOR LIGHTING TYPES—There are two basic types of outdoor lighting: Line-voltage lighting and low-voltage lighting. Line-voltage lighting operates on 120 volts, the same voltage used by lights and receptacles in the house, while low-voltage lighting operates on only 12 volts. Which type to use depends on a variety of factors including the size of the area to be lit, the purpose of the lighting, the desired flexibility of the system, and the construction budget. Each system has its advantages and disadvantages.

Hybrid systems, which take advantage of the strengths of both line- and low-voltage, are common. For example, line-voltage security lighting is often combined with low-voltage path lighting in the garden.

INSTALLATION—Landscape lighting is generally installed after the site has been finish-graded and the large plants put in. If planned for during construction, sleeves for the passage wires will have been provided below paved surfaces and in masonry structures such as walls. The work can be executed by a licensed electrician (120-volt systems), a landscape contractor (12-volt systems), or even a knowledgeable homeowner. Trenching for the wires may be done by anyone, because there are no legal requirements (other than depth) concerning the trenching.

Landscape lighting also can be worked into an existing site, though it can be an involved and costly project, depending on whether or not paved areas block the route of the system.

POWER SUPPLY—An electrical panel or subpanel mounted either inside or outside the house supplies power to both line- and low-voltage lighting systems. Because residential outdoor lighting tends to have minimal power requirements, the electrical circuits usually originate in panels that serve other electrical needs. In cold northern climates, the panels tend to be located inside the house, whereas in warmer climates, the panel is usually outdoors.

In the case of a line-voltage system, the cable that supplies the system is connected directly to a circuit breaker in the panel in the same manner as other circuits in the house. For a low-voltage system, the 120-volt electrical cable is run to a transformer that steps down the power to 12 volts AC.

Electrical panel—When the electrical panel is inside the building, the wires that power the landscape lighting must be routed from the building to the yard. For a line-voltage system, the transition from house to yard must comply with the electrical code and should be planned carefully to avoid extra work and cost (see 66A & 67A).

When the electrical panel is located outside the house, it is a simple matter to route the electrical wires for landscape lighting through a conduit to an underground location from where they can be run to the landscape lighting (see 67B).

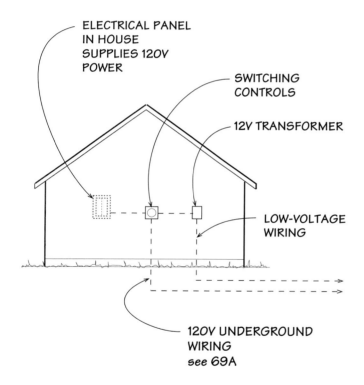

ELECTRICAL PANEL
IN HOUSE
SUPPLIES 120V
POWER

SWITCHING
CONTROLS

12V TRANSFORMER

LOW-VOLTAGE
WIRING

120V UNDERGROUND
WIRING
see 69A

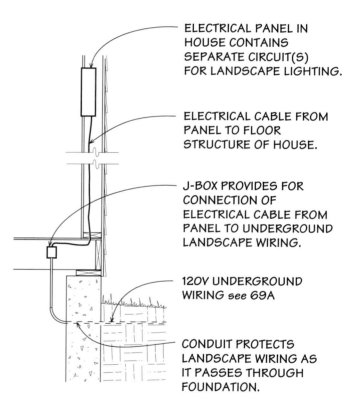

ELECTRICAL PANEL IN
HOUSE CONTAINS
SEPARATE CIRCUIT(S)
FOR LANDSCAPE LIGHTING.

ELECTRICAL CABLE FROM
PANEL TO FLOOR
STRUCTURE OF HOUSE.

J-BOX PROVIDES FOR
CONNECTION OF
ELECTRICAL CABLE FROM
PANEL TO UNDERGROUND
LANDSCAPE WIRING.

120V UNDERGROUND
WIRING see 69A

CONDUIT PROTECTS
LANDSCAPE WIRING AS
IT PASSES THROUGH
FOUNDATION.

(A) **120V POWER SUPPLY**
INSIDE PANEL/WOOD FLOOR

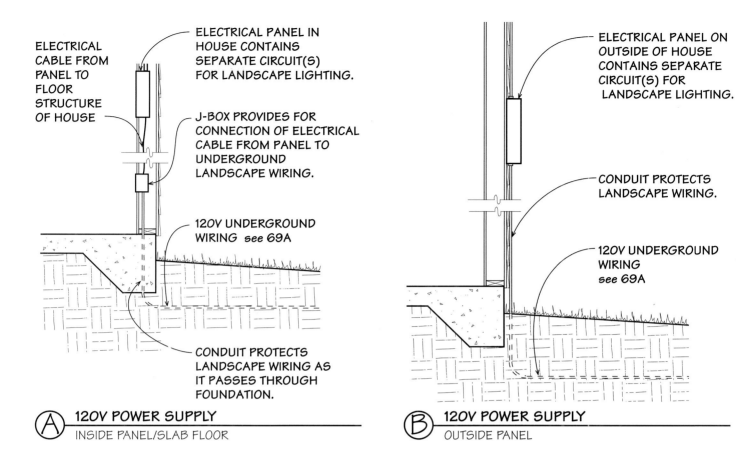

ELECTRICAL CABLE FROM PANEL TO FLOOR STRUCTURE OF HOUSE

ELECTRICAL PANEL IN HOUSE CONTAINS SEPARATE CIRCUIT(S) FOR LANDSCAPE LIGHTING.

J-BOX PROVIDES FOR CONNECTION OF ELECTRICAL CABLE FROM PANEL TO UNDERGROUND LANDSCAPE WIRING.

120V UNDERGROUND WIRING see 69A

CONDUIT PROTECTS LANDSCAPE WIRING AS IT PASSES THROUGH FOUNDATION.

Ⓐ **120V POWER SUPPLY**
INSIDE PANEL/SLAB FLOOR

ELECTRICAL PANEL ON OUTSIDE OF HOUSE CONTAINS SEPARATE CIRCUIT(S) FOR LANDSCAPE LIGHTING.

CONDUIT PROTECTS LANDSCAPE WIRING.

120V UNDERGROUND WIRING see 69A

Ⓑ **120V POWER SUPPLY**
OUTSIDE PANEL

Transformers—In low-voltage systems the transformer may be located either inside or outside the house. For indoor transformers, finding a path to the outside for the small-diameter, low-voltage wires is relatively easy and not regulated by codes.

Outdoor transformers must be of weather-tight construction, and the line-voltage cable that powers them must follow all code regulations. Outdoor transformers have the advantage of easy access by service personnel.

Hybrid systems can be created by locating a low-voltage system with its transformer anywhere within a line-voltage system.

WIRES AND CONDUIT—All electrical wires and cables for landscape wiring must be rated for their particular use.

Wires are rated according to their coatings and insulative jackets. Direct-burial cable is used for line-voltage landscape wiring in most residential applications. Wires are also rated for direct exposure to sunlight, for use in conduit, and for underwater use.

Wire—Wire size is described by an archaic inverse numbering system called gauge, which is related to the diameter of the wire—the larger the wire, the smaller the gauge. The selection of wire size depends on voltage, power (watts), and distance. For example, a line-voltage 200-watt light fixture can operate at the end of a 685-ft., #12 wire. The same wire for the same 200-watt load in a low-voltage system can only be approximately 35 ft. to 40 ft. long.

ACTUAL WIRE GAUGE SIZES

#18 & #16 ARE USED FOR LOW-VOLTAGE LIGHTING.
#14 & #12 ARE USED FOR LINE-VOLTAGE LIGHTING.

The extreme difference between the two examples is due to a phenomenon called voltage drop, which is a gradual decrease in power as the distance from the power source is increased. Voltage drop is much more significant in low-voltage systems, which is why line-voltage is recommended for systems involving great or even moderate distances.

Conduit—Conduit (also called a "raceway" in the code) is used to protect outdoor line-voltage wiring and is made either of metal or plastic (gray PVC per code requirements). Metal is preferred above ground because it is more durable, and plastic is preferred below grade because it does not corrode and is somewhat flexible, allowing it to follow gentle changes in contour and curves in trenching.

Thin-walled, white plastic piping is sometimes used to protect low-voltage wiring, even though protection is not required by code.

POWER USE (WATTS)	WIRE LENGTH ALLOWED (IN FEET)		
	#14 wire	#12 wire	#10 wire
200	410	685	1095
500	165	275	440
1200	70	115	185
1800	45	75	120

Line-Voltage Wire Size

Burial regulations—Underground line-voltage wiring is regulated by code and must be buried to a specified depth, which depends upon the amount of protection it has been given (see 69A). Direct-burial cable may be protected at a shallow depth by a concrete slab, or it may be buried very deep (24 in.) without physical protection. Direct-burial cable also may be located at an intermediate depth if protected by conduit or connected to a ground-fault current interrupter (GFCI), a device with a hair trigger designed to shut off the electricity at the slightest sign of trouble.

Low-voltage wiring is not dangerous, so it does not need to be buried for safety reasons, but it is usually buried at least slightly to protect it from garden tools.

CONNECTORS—Connections of line-voltage wiring for the purposes of branching the circuit or for connecting to fixtures, outlets, and switches are made in a metal or plastic junction box (J box).

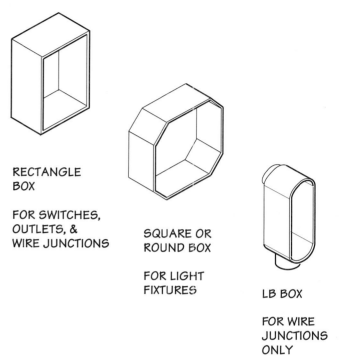

RECTANGLE BOX

FOR SWITCHES, OUTLETS, & WIRE JUNCTIONS

SQUARE OR ROUND BOX

FOR LIGHT FIXTURES

LB BOX

FOR WIRE JUNCTIONS ONLY

Boxes are designed for both above-grade and below-grade use and are usually connected to conduit, although they may be connected directly to electrical cable in some cases. Above-grade boxes are either weatherproof (WP) or standard, the latter being appropriate only when protected by a roof (see 70A to 71A).

For low-voltage connections, a variety of weather-tight connectors are made for above- or below-grade applications. These generally are designed to encase the spliced wires in grease or other non-water-soluble substances contained in a plastic casing. Low-voltage connections are also often made simply with a wire nut that is protected within the fixture (see 71B).

FINISH GRADE

2 IN.

DIRECT-BURIAL CABLE COVERED BY 2 IN. OF CONCRETE

12 IN.

DIRECT-BURIAL CABLE PROTECTED BY GROUND FAULT CURRENT INTERRUPTER (GFCI) @ PANEL

18 IN.

WIRES OR CABLE IN CONDUIT

24 IN.

DIRECT-BURIAL CABLE

NOTE: LOW-VOLTAGE WIRE HAS NO BURIAL REQUIREMENTS, BUT SHOULD BE BURIED TO PROTECT AGAINST LAWN EDGERS & OTHER GARDEN TOOLS.

 LINE VOLTAGE BURIAL DEPTH

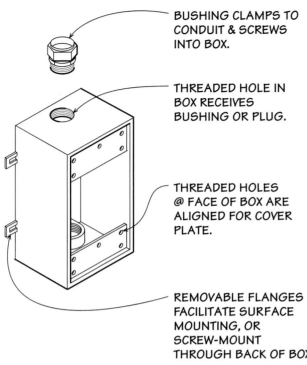

BUSHING CLAMPS TO CONDUIT & SCREWS INTO BOX.

THREADED HOLE IN BOX RECEIVES BUSHING OR PLUG.

THREADED HOLES @ FACE OF BOX ARE ALIGNED FOR COVER PLATE.

REMOVABLE FLANGES FACILITATE SURFACE MOUNTING, OR SCREW-MOUNT THROUGH BACK OF BOX.

(A) 120V WP J-BOX

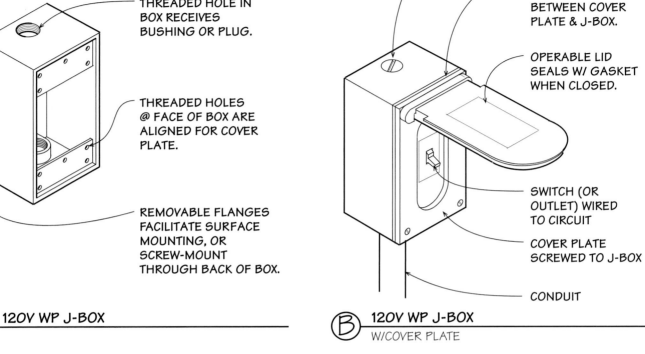

SCREW IN PLUG & SEAL WHEN NO CONDUIT IN PLACE.

COMPRESSIBLE GASKET SEALS BETWEEN COVER PLATE & J-BOX.

OPERABLE LID SEALS W/ GASKET WHEN CLOSED.

SWITCH (OR OUTLET) WIRED TO CIRCUIT

COVER PLATE SCREWED TO J-BOX

CONDUIT

(B) 120V WP J-BOX
W/COVER PLATE

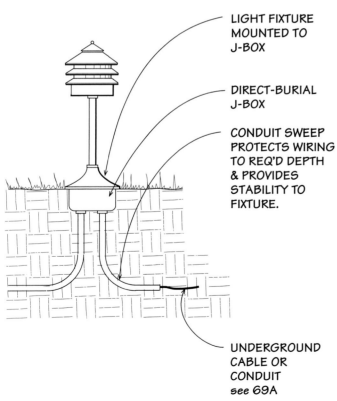

LIGHT FIXTURE MOUNTED TO J-BOX

DIRECT-BURIAL J-BOX

CONDUIT SWEEP PROTECTS WIRING TO REQ'D DEPTH & PROVIDES STABILITY TO FIXTURE.

UNDERGROUND CABLE OR CONDUIT
see 69A

(C) 120V DIRECT-BURIAL J-BOX

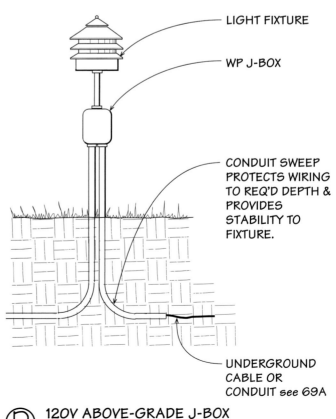

LIGHT FIXTURE

WP J-BOX

CONDUIT SWEEP PROTECTS WIRING TO REQ'D DEPTH & PROVIDES STABILITY TO FIXTURE.

UNDERGROUND CABLE OR CONDUIT see 69A

(D) 120V ABOVE-GRADE J-BOX

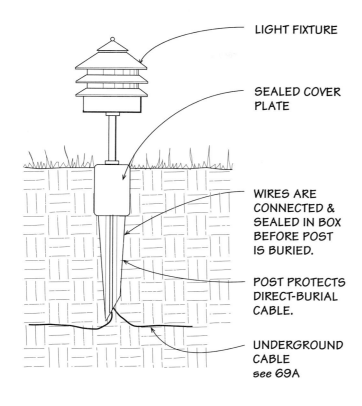

LIGHT FIXTURE

SEALED COVER PLATE

WIRES ARE CONNECTED & SEALED IN BOX BEFORE POST IS BURIED.

POST PROTECTS DIRECT-BURIAL CABLE.

UNDERGROUND CABLE
see 69A

 120V DIRECT-BURIAL POST BOX

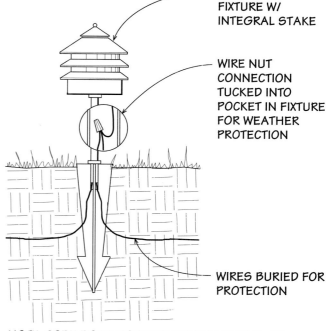

LOW-VOLTAGE FIXTURE W/ INTEGRAL STAKE

WIRE NUT CONNECTION TUCKED INTO POCKET IN FIXTURE FOR WEATHER PROTECTION

WIRES BURIED FOR PROTECTION

NOTE: CODE DOES NOT REQUIRE THE BURIAL OF LOW-VOLTAGE WIRES, BUT THEY ARE OFTEN BURIED &/OR PLACED IN LOW-COST IRRIGATION PIPE FOR PROTECTION.

LOW-VOLTAGE FIXTURE

LIGHT FIXTURES—The catalogs of available residential-landscape lighting fixtures fill volumes. Fixtures are designed for general illumination, accent lighting, security lighting, path lighting, and a host of other specific tasks. There are fixtures designed to mount in the ground, on walls, on ceilings, on trees, and under water (see 72A to 75B).

Experienced landscape lighting designers are familiar with all types of fixtures and with the principles of their use, and their consultation can add value to most landscape lighting projects. A basic criterion for good lighting design is not being able to see the actual light source (i.e., the bulb). Much like a lampshade on an indoor fixture, masking the outdoor light source will focus attention on the area or element being lit rather on the light itself.

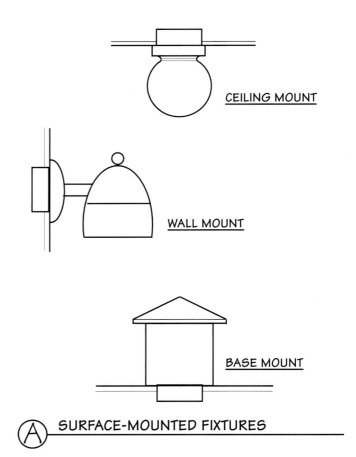

CEILING MOUNT

WALL MOUNT

BASE MOUNT

Ⓐ SURFACE-MOUNTED FIXTURES

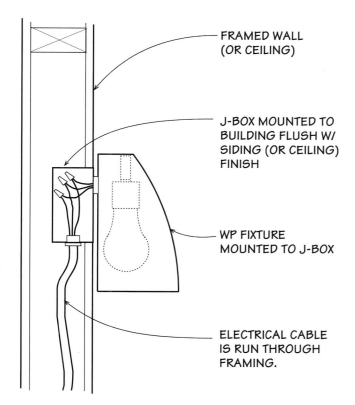

FRAMED WALL
(OR CEILING)

J-BOX MOUNTED TO
BUILDING FLUSH W/
SIDING (OR CEILING)
FINISH

WP FIXTURE
MOUNTED TO J-BOX

ELECTRICAL CABLE
IS RUN THROUGH
FRAMING.

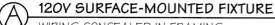

(A) 120V SURFACE-MOUNTED FIXTURE
WIRING CONCEALED IN FRAMING

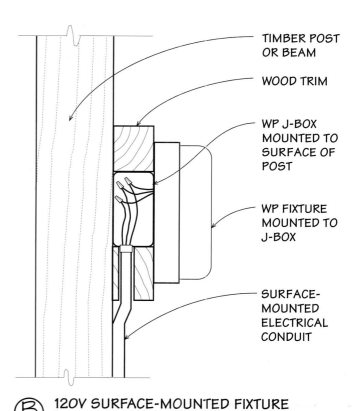

TIMBER POST
OR BEAM

WOOD TRIM

WP J-BOX
MOUNTED TO
SURFACE OF
POST

WP FIXTURE
MOUNTED TO
J-BOX

SURFACE-
MOUNTED
ELECTRICAL
CONDUIT

(B) 120V SURFACE-MOUNTED FIXTURE
WIRING IN CONDUIT

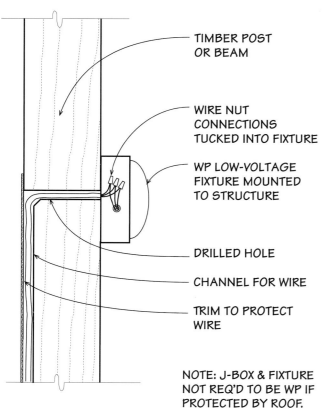

TIMBER POST
OR BEAM

WIRE NUT
CONNECTIONS
TUCKED INTO FIXTURE

WP LOW-VOLTAGE
FIXTURE MOUNTED
TO STRUCTURE

DRILLED HOLE

CHANNEL FOR WIRE

TRIM TO PROTECT
WIRE

NOTE: J-BOX & FIXTURE
NOT REQ'D TO BE WP IF
PROTECTED BY ROOF.

(C) 12V SURFACE-MOUNTED FIXTURE
WIRING CONCEALED IN TIMBER

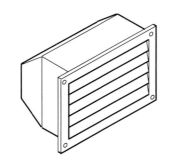

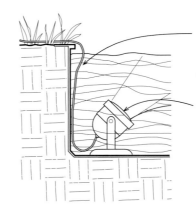

CORD EXTENDS OUT OF WATER TO 120V GFCI OUTLET OR TO 12V TRANSFORMER.

120V OR 12V FREESTANDING, SEALED UNDERWATER FIXTURE RESTS ON BOTTOM OF POND.

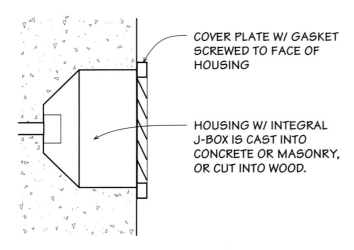

COVER PLATE W/ GASKET SCREWED TO FACE OF HOUSING

HOUSING W/ INTEGRAL J-BOX IS CAST INTO CONCRETE OR MASONRY, OR CUT INTO WOOD.

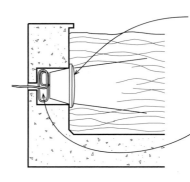

120V SEALED UNDERWATER FIXTURE MOUNTS IN HOUSING CAST INTO CONCRETE POOL WALL.

CORD COILED @ REAR OF HOUSING ALLOWS FIXTURE TO BE REMOVED FROM WATER FOR BULB REPLACEMENT.

Ⓐ RECESSED FIXTURE

Ⓑ SUBMERGED FIXTURES

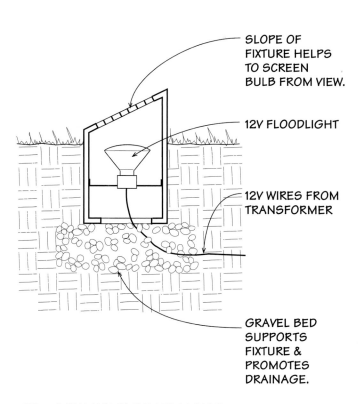

SLOPE OF FIXTURE HELPS TO SCREEN BULB FROM VIEW.

12V FLOODLIGHT

12V WIRES FROM TRANSFORMER

GRAVEL BED SUPPORTS FIXTURE & PROMOTES DRAINAGE.

 A LOW-VOLTAGE UP-LIGHT

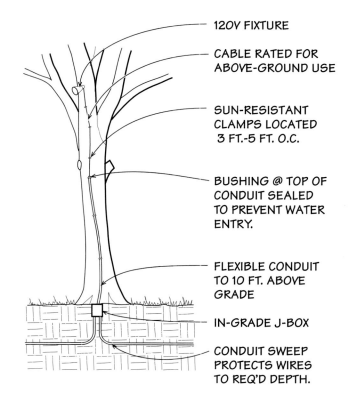

120V FIXTURE

CABLE RATED FOR ABOVE-GROUND USE

SUN-RESISTANT CLAMPS LOCATED 3 FT.-5 FT. O.C.

BUSHING @ TOP OF CONDUIT SEALED TO PREVENT WATER ENTRY.

FLEXIBLE CONDUIT TO 10 FT. ABOVE GRADE

IN-GRADE J-BOX

CONDUIT SWEEP PROTECTS WIRES TO REQ'D DEPTH.

B 120V TREE-MOUNTED FIXTURE

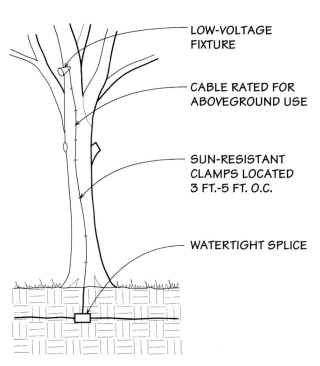

LOW-VOLTAGE FIXTURE

CABLE RATED FOR ABOVEGROUND USE

SUN-RESISTANT CLAMPS LOCATED 3 FT.-5 FT. O.C.

WATERTIGHT SPLICE

 C 12V TREE-MOUNTED FIXTURE

OUTLETS AND SWITCHES—Electrical outlets and switches located in the yard often add convenience and utility to a landscape lighting system. Outlets can be useful at an outdoor barbecue or for plugging in garden equipment, and switches at gazebos or other outdoor gathering areas are quite practical. Outlets and switches are mounted in weatherproof junction boxes and tied to the lighting system through connecting wires.

LIGHTING CONTROL—There are many ways to control a landscape lighting system. Control devices operate by interrupting the positively charged (hot) wire in the 120-volt cable that supplies power to the lights.

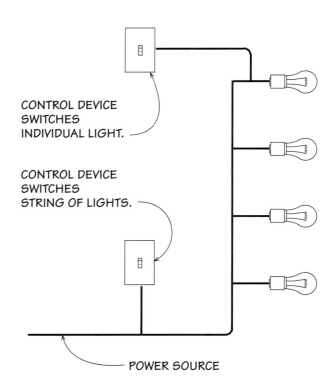

CONTROL DEVICE SWITCHES INDIVIDUAL LIGHT.

CONTROL DEVICE SWITCHES STRING OF LIGHTS.

POWER SOURCE

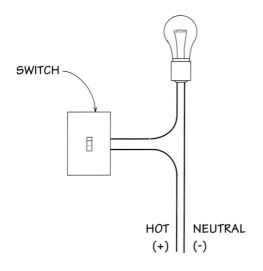

SWITCH

HOT
(+)

NEUTRAL
(-)

Typically, groups of lights are switched together, but any of the control devices may also be used for an individual light.

A programmable automatic timer (mechanical or electronic) can be substituted for the simple on/off switch for convenience. Lights also can be controlled by a photoelectric cell that activates the lighting when night falls or by a motion detector that turns on lights

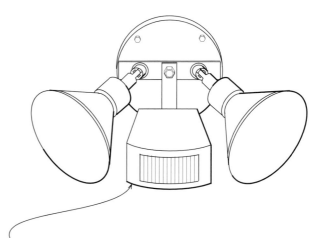

MOTION DETECTOR (OR PHOTOELECTRIC CELL) CAN BE MOUNTED ON SAME FIXTURE W/ LIGHTS OR IT CAN BE MOUNTED INDEPENDENTLY. THESE 120V DEVICES CAN CONTROL SINGLE OR MULTIPLE LIGHTS.

 MOTION DETECTOR

when an infrared beam is interrupted (see 76A). Devices can be used in combination for increased control. For example, a photoelectric cell can be used to turn on a circuit at dusk, while an automatic timer turns it off at the appropriate time.

Low-voltage lighting is typically controlled by means of the 120-volt cable that supplies the transformer rather than from the 12-volt line itself. For this reason, low-voltage lights are almost always switched as a group. Some manufacturers of low-voltage systems now offer a remote control that will turn a transformer on or off.

WIRING ON OPEN STRUCTURES—Anyone with an eye for aesthetics has noticed the difficulty of elegantly mounting conduit and junction boxes on the surface of structures. Yet conduit and junction boxes cannot be avoided on open structures such as gazebos and trellises if line-voltage electrical lighting is to be incorporated. The conduit usually appears to be an afterthought even if designed integrally with the structure because it occurs only sporadically where wires must be run, and it must make sweeping bends that rarely correspond with the dimensions of the structure (see 77A). Junction boxes also must be mounted wherever an electrical device appears.

Where appearance is a serious concern, the most common and practical approach for line-voltage lighting is to attempt to hide the conduit. It may be run inside of hollow columns. It can be hidden from view on the tops of beams. It can be painted the same color as the structure. The best alternative, however, is to use low-voltage wiring on open structures so that the conduit can be avoided altogether. The thin, low-voltage wires can be run on the surface or beneath trim, and they can make sharp bends, remaining concealed from view while moving from one member to another.

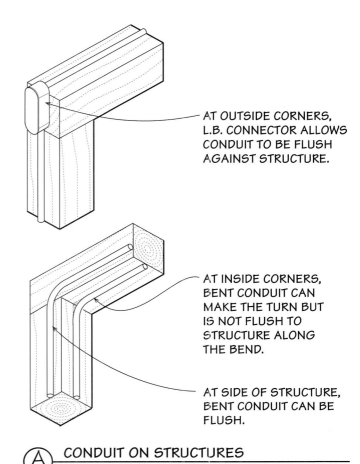

AT OUTSIDE CORNERS, L.B. CONNECTOR ALLOWS CONDUIT TO BE FLUSH AGAINST STRUCTURE.

AT INSIDE CORNERS, BENT CONDUIT CAN MAKE THE TURN BUT IS NOT FLUSH TO STRUCTURE ALONG THE BEND.

AT SIDE OF STRUCTURE, BENT CONDUIT CAN BE FLUSH.

(A) CONDUIT ON STRUCTURES

CHAPTER 3

PAVING

Like the floors inside a building, the paved areas outside a structure support myriad activities, from parking a car to playing basketball, riding bikes, eating, or socializing. These areas in the landscape must be designed to meet the demands that various uses will place upon them. Their design can go far beyond functional, however, providing the landscape architect or designer with a diverse palette of materials and options that can enhance the aesthetic qualities of a landscape while also addressing accessibility, sustainability, and durability.

SITE CONSIDERATIONS

Before deciding on a specific type of paving material for a given area, a number of issues with respect to the site and its intended uses must be considered. To ensure that the paving will meet the homeowner's functional and aesthetic demands, the following questions should be considered:

• What outdoor activities need a paved surface?

• Is the area to be paved a single-use space (such as a garden path or utility area), or is it an area that will serve multiple purposes (such as a driveway that must accommodate cars as well as bike riding, basketball, or other games)?

• Is it an area of high use or high traffic, or is it a lightly used space? Will traffic be in the form of foot traffic, or will it include wheeled traffic as well? Are heavy vehicles likely to travel on it?

• Should the paving be rigid (like concrete), semi-rigid (such as brick pavers), or loose (such as compacted gravel or decomposed granite)?

• Is permeable paving (paving that allows storm water to penetrate into the subsurface) an option to consider, or is impermeable paving the preferred option (due to soils that drain poorly, for example)?

• Is weather a consideration? Is this surface likely to need shoveling or plowing to remove snow? Are summer temperatures extreme? Is moss buildup from long, wet periods a concern?

• What aesthetic look is desired—formal and highly maintained, or informal and more cottagelike?

• Where are the paved areas in relation to other elements (such as buildings, trees, or streets), and how might the materials in those elements influence the choice of paving material? For example, is a flagstone path the best choice aesthetically if the house is made of brick?

• What level of maintenance is reasonable or required, given the nature of the paved space, its placement on the site (front yard versus back, for example), and the site-specific and regional impacts of things such as vegetation, weather, or temperature?

• How flexible is the budget for this portion of the project?

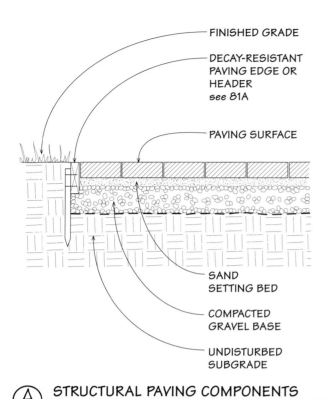

FINISHED GRADE

DECAY-RESISTANT
PAVING EDGE OR
HEADER
see 81A

PAVING SURFACE

SAND
SETTING BED

COMPACTED
GRAVEL BASE

UNDISTURBED
SUBGRADE

(A) STRUCTURAL PAVING COMPONENTS

PAVING BASICS

Whether selecting a rigid or loose paving system, you will find a number of terms used to describe the structural components of the systems are common to all or many paving choices (see 80A).

SUBGRADE—Subgrade refers to the soil upon which all of the material (subbase, paving material) is set. This is typically the existing soil that is left after excavation or that is placed and compacted as fill (see Chapter 1, p. 13). If the subgrade is undisturbed soil (not fill), then it should need little or no compaction prior to paving installation; if it is fill, then it should be compacted with a mechanical, plate-type compactor to ensure that the soil will not settle over time. Subgrade is typically compacted to 95 percent, which is nearly as compacted as undisturbed soil.

Most native soils will be acceptable as a subgrade, although those with a high level of organic matter (such as topsoil or compost) or clay content (which is susceptible to expansion and contraction due to changes in ground moisture) should be excavated and replaced with a well-drained subgrade material.

SUBBASE—Subbase is the layer between the subgrade and the paving material. It frequently sits directly on the subgrade, but sometimes a fabric barrier (referred to as landscape fabric or filter fabric) is used between the subgrade and subbase.

The subbase is typically a well-drained, crushed gravel containing "fines" (the rock dust that is created when larger rocks are ground and crushed) that is easily compacted. It provides structural strength to the paving system by transferring the weight of the loads placed on the paving surface to the subgrade below. This allows the paving surface to maintain its shape and condition over time.

The type and dimension of the gravel subbase varies from region to region, but the gravel is generally around ¾ in. in diameter and smaller (typically called ¾ minus), and the subbase depth ranges from 4 in. to 6 in., or greater in regions where frost and freezing occur. Check with local contractors or suppliers for recommendations specific to your area.

PAVING MATERIAL—This is the top layer of the paving system. The paving material itself may be composed of several levels, however, such as concrete pavers that require a sand "setting bed," or asphalt paving, which is typically poured in two separate "courses" (a base course set upon the subbase and a wearing course upon which traffic moves). There is a wide range of paving materials to choose from, including precast concrete pavers, stone, brick, tile, gravel and other granular materials, and several specialized surfaces, such as rubberized surfaces for play areas.

PAVING EDGE—Many paving systems, particularly those that are not rigid, require a rigid edge to keep the paving material in place. This edge can be made of many different materials, ranging from wood and plastic to steel and concrete; some paving manufacturers make special edging material for use in conjunction with a specific paving material (see 81A). Sometimes referred to as a "header," or even as a "mowstrip" (if it occurs where paving meets grass and is wide and strong enough to support a lawnmower), the presence of a well-designed edge can ensure that the paving system will maintain its shape and condition over time. One important aspect of a well-designed edge is its ability to remain in place without flexing over time. Plastic garden edging, for example,

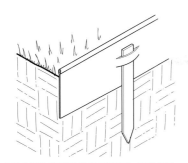

PLASTIC EDGING W/ STEEL STAKE

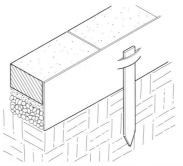

METAL EDGING W/ STEEL STAKE

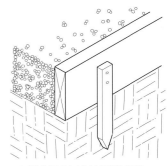

WOODEN HEADER W/
DECAY-RESISTANT STAKE

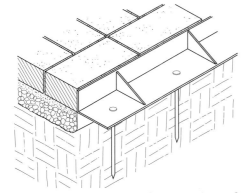

PLASTIC "SNAP EDGE" W/ STEEL PIN ANCHOR

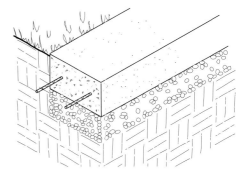

CONCRETE "MOWSTRIP" W/ CONTINUOUS REBAR

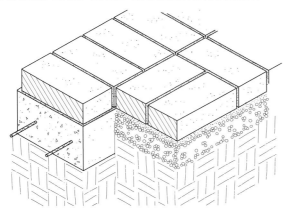

MORTARED BRICK ON CONCRETE FOUNDATION
(SHOWN ADJACENT TO BRICK ON SAND & GRAVEL BASE)

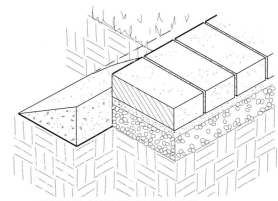

BURIED CONCRETE HEADER
(SHOWN ADJACENT TO PAVERS ON SAND
& GRAVEL BASE)

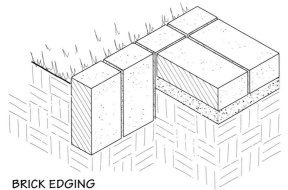

BRICK EDGING
(NOT MORTARED; SHOWN ADJACENT TO BRICK ON SAND SETTING BED)

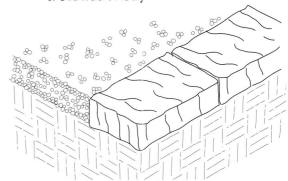

STONE-IN-SOIL EDGING (SHOWN
ADJACENT TO COMPACTED GRAVEL PATH)

 TYPES OF PAVING EDGES

TYPES OF PAVING EDGES

Type of material	Comments
Plastic flexible edging	Not recommended for paving applications due to its lack of strength; allows significant lateral movement w/ almost all paving materials.
Metal edging (aluminum, galvanized, or painted steel)	Very strong, yet flexible. Good for curvilinear pathways or for paving systems where a minimally visible edge is desired. Typical use is along brick paths, concrete paver pathways, & stone walks. Held in place w/ metal stakes.
Wood (decay-resistant)	Less expensive option but shorter lived. Even decay-resistant materials will lose their strength after a relatively short period. Good for less formal areas & as a separation between planting areas & paved areas that utilize semirigid or loose paving systems. Best for straight runs, although curves are possible w/ smaller-dimensioned wood. Typically anchored w/ wooden, decay-resistant stakes placed every 3 ft. to 4 ft.
Plastic "Snap Edge"	Prefabricated material, sold in conjunction w/ paving systems such as concrete pavers. Creates a stable, subgrade edge that minimizes lateral movement. Typ. use is w/ paving systems like concrete pavers. Not suitable for use w/ gravel or asphalt because it is intended to be placed below the surface level of the paving material. Good in both straight & curvilinear applications.
Concrete Visible "mowstrip"	Very long-lasting & durable, good for edges along grass areas where mowers & edgers are run. Creates a very visible line in the landscape, although coloring of the concrete can lessen visible impact. Expensive to install & should incorporate one or two linear, continuous runs of rebar (#3 or #4) along entire length of mowstrip. Also can be used to border paved areas where different materials are used w/in the borders of concrete strip, such as pavers, brick, or loose paving materials such as gravel. Good in both straight & curvilinear applications.
Not visible, below-grade header	Used in situations similar to those described in Plastic Snap Edge section. Good for situations where an invisible header is desired. Effective method for minimizing lateral movement of paving material; this is enhanced w/ the addition of a continuous run of rebar along the entire length of the header.

is not a good choice for use as a paving edge because it bends and distorts fairly easily, which would allow paving material to shift and move along with it.

PAVING SYSTEMS

When selecting a paving system, the most important facet to examine is the level of activity and traffic, or wear and tear, that the paving will need to hold up under. Different paving systems have different levels of long-term maintenance, involving varying degrees of cost and difficulty, which also should be considered during the selection process.

TYPES OF PAVING EDGES

Type of material	Comments
Brick Mortared	Similar to a concrete mowstrip in its use, it is most commonly seen as an edge material in conjunction w/ brick-on-sand paving or as an accent strip along a poured-in-place concrete walk. Also can be an effective & attractive edge along a less formal gravel walk. Brick is mortared in place on top of a concrete footer, typ. in either a single brick wide, side-by-side, or basket-weave pattern. Concrete footer should incorporate rebar along its entire length (#3 or #4 continuous). Other materials, such as stone, also can be mortared on top of a concrete strip in much the same manner. Good in both straight & curvilinear applications.
Brick Not mortared	Bricks are placed vertically in a trench w/ the ends leveled & soil compacted tightly to minimize lateral movement. Good for use w/ informal brick & gravel paths. Good in both straight & curvilinear applications, although aesthetically generally best suited to curved-path applications. Do not use in high-traffic areas or in soils that have a high-sand or organic-matter content because edging will shift significantly. Movement of material also may create a tripping hazard—maintenance is key.
Stone (not mortared)	Similar in some respect to the unmortared brick edging described above. Best used in informal situations, particularly in conjunction w/ loose paving materials such as compacted gravel. Stones should be larger than 12 x 12 in. & thicker than 2 in. Stones are installed directly into compacted soil & are then leveled & set by compacting soil around the rock's edges. Not recommended for high-traffic areas. Rock will shift over time & may need to be reset. Do not use in soils w/ a high sand or organic matter content.

This is not intended to be an exhaustive list. Check w/ local contractors or design professionals for additional options that may be specific to your region.

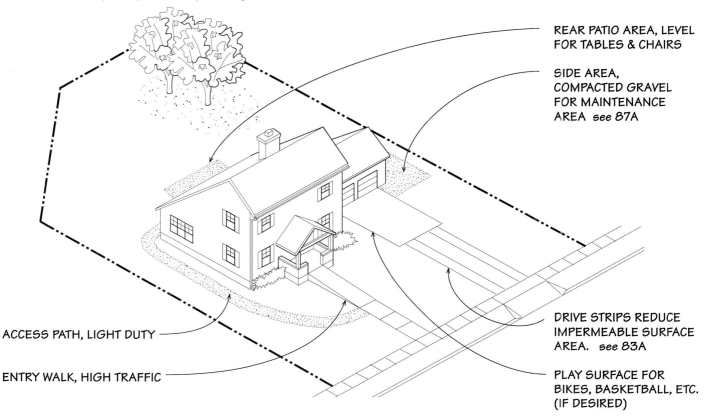

REAR PATIO AREA, LEVEL FOR TABLES & CHAIRS

SIDE AREA, COMPACTED GRAVEL FOR MAINTENANCE AREA see 87A

DRIVE STRIPS REDUCE IMPERMEABLE SURFACE AREA. see 83A

PLAY SURFACE FOR BIKES, BASKETBALL, ETC. (IF DESIRED)

ACCESS PATH, LIGHT DUTY

ENTRY WALK, HIGH TRAFFIC

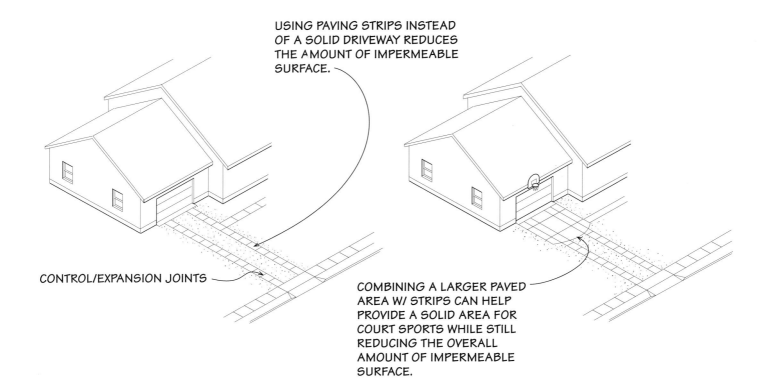

USING PAVING STRIPS INSTEAD OF A SOLID DRIVEWAY REDUCES THE AMOUNT OF IMPERMEABLE SURFACE.

CONTROL/EXPANSION JOINTS

COMBINING A LARGER PAVED AREA W/ STRIPS CAN HELP PROVIDE A SOLID AREA FOR COURT SPORTS WHILE STILL REDUCING THE OVERALL AMOUNT OF IMPERMEABLE SURFACE.

RIGID AND SEMIRIGID SYSTEMS—In higher-traffic areas, or in areas where a more "kept" appearance is desired, rigid and semirigid paving are probably the best choices. Paving types that fall under this category include concrete, pavers, mortared tile or brick, stone, and asphalt. Once installed, the configuration of these materials will not change (or will change very little, as in the case of pavers or brick installed on a sand base) as people and vehicles move over the surface.

Rigid and semirigid paving materials tend to be more expensive per sq. ft. to purchase and install than nonrigid materials, but they require less maintenance over the life of the product and hold up better in high-traffic areas, such as driveways, patios, and primary pathways. They also support multiple uses more easily, especially if the uses are high-intensity, such as basketball, bike riding, skateboarding, or car parking.

NONRIGID SYSTEMS—Loose, or nonrigid, paving materials include crushed rock (sometimes called quarter minus, which is a gravel mix comprising gravel pieces ¼ in. in diameter and below with fines included), pea gravel (small-diameter rock that is round, not crushed, and free of fines), decomposed granite, flagstones, and several other materials that vary from region to region (such as crushed washed shells in parts of the Southeast). These materials will shift by varying degrees over time and will need a higher level of maintenance to stay in good condition. In some instances, they also will need a rigid edge to prevent them from expanding (also described as "squishing out") into the soil adjacent to the paved area.

Preferred Paving Materials from Most Expensive to Least Expensive (per sq. ft.)*

Material	Estimated Direct Cost
Mortared tile/brick over concrete	$7–$15
Mortared stone over concrete	$8–$12
Cut stone on sand	$6–$15
Brick/concrete pavers on sand	$3.50–$5
Flagstone on sand	$3–$9
Concrete w/ special finish (color, stamped finish)	$5–$6
Asphalt (two lifts)	$3–$6
Concrete w/ typical finish	$1–$3
Decomposed granite (depends on thickness & availability)	$1–$4
Asphalt (one lift)	$1–$2
Compacted gravel w/ fines (¼ minus) (depends on thickness)	.50–$1.50
Fieldstone stepping-stones	varies greatly
Concrete stepping-stones	.50–$2

*Note: This list is based upon average costs. Actual costs may vary due to regional differences & shipping costs. There also may be paving types not listed, since many regional variations exist. Check w/ local suppliers to determine availability and actual costs.

Loose paving materials require less expertise to install and are cheaper per square foot, but they require more maintenance over time. They possess a less formal, more relaxed aesthetic, which can be good or bad, depending on what the client is trying to achieve, and they are also typically more permeable than rigid paving materials, which can be good or bad, depending on the surrounding soil.

MATERIALS AND FINISHES—When considering types of paving systems, it is important to think about the surface finish of a material. The wrong finish can negatively impact the comfort and usability of an area. Flagstone paving, for example, can be uncomfortable and difficult to walk on, while being nearly impossible to navigate with a wheelchair. It can also be difficult to set a table and chairs on flagstone unless a relatively smooth stone, such as slate, is used.

Conversely, finishes that are too smooth can cause traction problems, particularly in regions with heavy rain and snow. Concrete paving with a very light broom finish may be fine in the Southwest, for example, but a more textured, slip-resistant finish would be preferable in the Northwest. More information on specific textures, colors, and finishes appears later in this chapter.

WEATHER AND MAINTENANCE—You also should consider how well a paving system will hold up under regional and seasonal weather conditions. Flagstones, for example, create an elegant pathway, but their uneven surface can make it difficult to shovel in snowy climates. Pea gravel offers the benefits of easy installation and water percolation, but if it is placed near deciduous planting beds, it will be very difficult to rake the area without removing a significant amount of gravel along with the leaves—an ongoing maintenance headache.

UNDERGROUND ACCESS—Also consider the long-term implications of paving selection. The most important of these is the ease with which a paving system can be removed, repaired, or replaced. Both loose and semirigid paving systems, for example, allow for relatively easy access to underground utilities such as irrigation, sewer, and gas lines. Once underground work has been completed, the paving system can be relaid, leaving no visual indication of any work having been done and requiring no additional purchase of new material.

A surface paved with concrete, however, would need to be cut and broken out, and then replaced with new concrete. In most situations, this would leave a "scar" where the work was done. The same problem would arise if areas of the paved surface were to break down or fail, necessitating repair or replacement.

PERMEABLE PAVING

Chapter 1 introduced storm-water management as an important consideration in the design of a landscape. Paving can create an increased level of runoff into a city's storm-water system that will inevitably contribute negatively to the health of nearby streams and waterways. As a result, more attention has recently been paid to the development of paving systems that allow water to percolate down into the subgrade and into the aquifer below. These systems, classified as permeable, reduce the overland runoff leaving a site by channeling at least some of the storm-water down through the base material and into the soil below.

While a permeable system's efficiency primarily depends on the level of porosity that a specific paving material has, it also depends on several other elements. These include the capacity of both subbase and subgrade to facilitate the movement of water to the soil horizons below, the subgrade's level of compaction, and adequate surface maintenance to ensure that the paving material is open to water percolation. Even with an effective permeable system, however,

the paved surface should still be sloped a minimum of 1 percent (approximately 1 in. per 8 ft. to 10 ft.) in much the same manner as impermeable systems (see pp. 90–95).

Using permeable paving systems can benefit the homeowner by lowering sewer charges (some jurisdictions offer reduced rates for customers who use permeable paving systems) and construction costs associated with the expensive, piped drainage systems required to handle storm water. Areas with well-drained soils are prime candidates for permeable paving. In areas where soils are poorly drained, permeable paving can still be used, but they should be combined with the drainage systems described in chapter 1.

GRAVEL AND GRAVEL-LIKE MATERIALS—Garden paths, utility areas, and patios where walking and sitting are the primary uses are good areas for gravel paving. Gravel should be avoided in areas that support higher-impact activity like running because it doesn't offer much traction. It is also not ideal for use in areas with car traffic unless it is well contained at the edges and placed on a well-compacted and well-drained subbase. A fabric layer beneath the subbase can lessen the amount of settling that occurs from the weight of the vehicles.

Gravel used for paving is typically ¾-in. or less in diameter, and can either be rounded and smooth or fractured and rough. "Pea" or "sized" gravels are typically used in more decorative areas where traffic is minimal because they can't be compacted and tend to migrate, even with a rigid edge. Pea gravels do, however, allow better percolation than gravels containing fines.

For paving areas that will receive medium or high levels of traffic, gravels that have a mixture of small, crushed, angular rock and fines are the best choice. Similar materials, such as decomposed granite or crushed washed shells, also fall within this category. While the top surface will still migrate a bit, the material, once compacted, stays in place rather well with the help of a rigid edge (see 87A).

FLAGSTONE—Stone pathways and patios can be beautiful additions to a landscape while also providing the homeowner with a durable and functional surface. Flagstone is a generic term used to describe

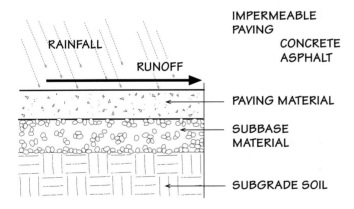

RAINFALL
RUNOFF
IMPERMEABLE PAVING
CONCRETE
ASPHALT
PAVING MATERIAL
SUBBASE MATERIAL
SUBGRADE SOIL

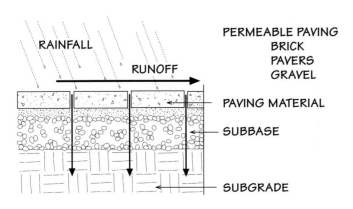

RAINFALL
RUNOFF
PERMEABLE PAVING
BRICK
PAVERS
GRAVEL
PAVING MATERIAL
SUBBASE
SUBGRADE

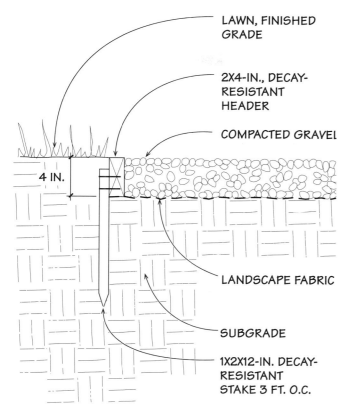

LAWN, FINISHED GRADE

2X4-IN., DECAY-RESISTANT HEADER

COMPACTED GRAVEL

4 IN.

LANDSCAPE FABRIC

SUBGRADE

1X2X12-IN. DECAY-RESISTANT STAKE 3 FT. O.C.

 GRAVEL PATH W/WOODEN HEADER

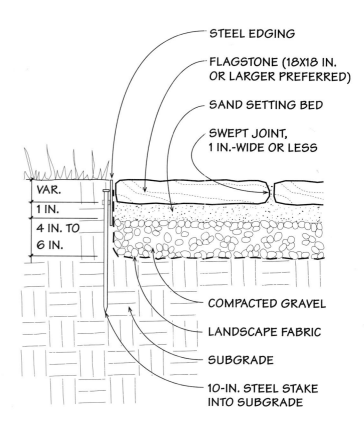

STEEL EDGING

FLAGSTONE (18X18 IN. OR LARGER PREFERRED)

SAND SETTING BED

SWEPT JOINT, 1 IN.-WIDE OR LESS

VAR.
1 IN.
4 IN. TO 6 IN.

COMPACTED GRAVEL

LANDSCAPE FABRIC

SUBGRADE

10-IN. STEEL STAKE INTO SUBGRADE

 FLAGSTONE PAVING

paving stones that are typically larger than 1 ft. to 1½ ft. square. Installation methods vary from region to region and also will be impacted by how and where they'll be used, and the desired aesthetic.

Flagstone paving can be installed directly into the soil (if the soil is structurally stable) or it can be set on a gravel bed and filled in with a soil mix that will support vegetation for a more naturalistic look. Flagstone also can be installed on a gravel subbase and sand setting bed (much like brick or concrete pavers), but care should be taken to ensure that the gaps between the stones are less than ½ in. to prevent rapid erosion; this application is also recommended for nonstructural soils such as those with a high clay or organic content (see 87B). Finally, flagstones can be set in mortar on a concrete slab (see impermeable paving systems beginning on p. 90), but this eliminates water's ability to percolate.

Setting flagstones directly into the ground works for light-duty pathways, but in areas with more traffic, a gravel subbase should be used along with a paving

edge (galvanized or finished steel edging is a good choice because it is sturdy but pliable).

PAVERS OR BRICK ON SAND—Precast concrete paving systems and clay brick, when installed on a sand setting bed, offer a wonderful surface that can support a range of uses. While similar in installation, they are very different in both aesthetics durability.

Brick on sand—Traditional brick is composed of kiln-fired clay. It is a durable material that weathers nicely over time, acquiring a more rustic patina than concrete pavers. Brick on sand is typically laid with fairly "tight" or narrow gaps in between each brick, which helps prevent movement over time (see 88B).

Occasionally, a sandy soil mix is swept into the joints and then planted with seed, or even with small groundcover plugs. While this increases the rustic appearance, it makes maintenance a bit more difficult and decreases the ability of the surface to support wheeled traffic.

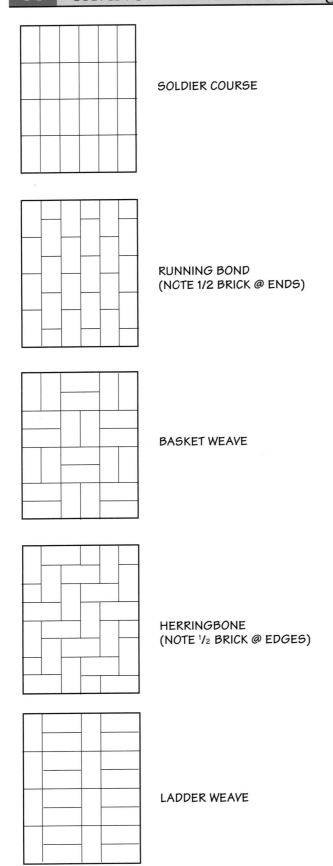

SOLDIER COURSE

RUNNING BOND
(NOTE 1/2 BRICK @ ENDS)

BASKET WEAVE

HERRINGBONE
(NOTE 1/2 BRICK @ EDGES)

LADDER WEAVE

Ⓐ BRICK PATTERNS

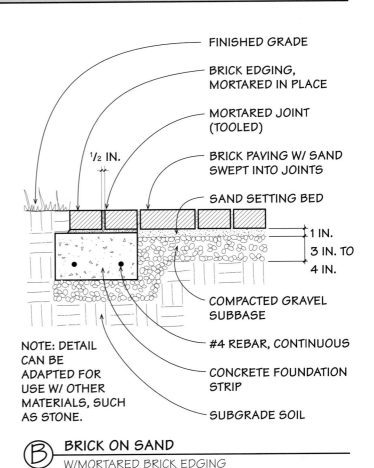

FINISHED GRADE

BRICK EDGING,
MORTARED IN PLACE

MORTARED JOINT
(TOOLED)

1/2 IN.

BRICK PAVING W/ SAND
SWEPT INTO JOINTS

SAND SETTING BED

1 IN.

3 IN. TO
4 IN.

COMPACTED GRAVEL
SUBBASE

NOTE: DETAIL
CAN BE
ADAPTED FOR
USE W/ OTHER
MATERIALS, SUCH
AS STONE.

#4 REBAR, CONTINUOUS

CONCRETE FOUNDATION
STRIP

SUBGRADE SOIL

Ⓑ BRICK ON SAND
W/MORTARED BRICK EDGING

There are a number of different patterns that are used when installing brick on sand. The most popular are running bond, soldier course, basket weave, and herringbone. (see 88A) When selecting a pattern, consider the shape of the pathway itself, because some patterns work well with curves (running bond), while others are better suited to straight, or more rectilinear areas (basket weave). In all cases, rigid paving edge should be used to prevent migration, which can create unsightly (and dangerous) gaps and bumps.

Precast concrete pavers—Pavers are a wonderful material to use in high-traffic situations, including driveways. Their ability to withstand heavy weight while allowing some percolation (though not as much as gravel, flagstone, or brick) makes them an attractive alternative to impermeable paving (see 89B). Some types of pavers, often referred to as eco-pavers, are specifically designed to enhance percolation (see 89A).

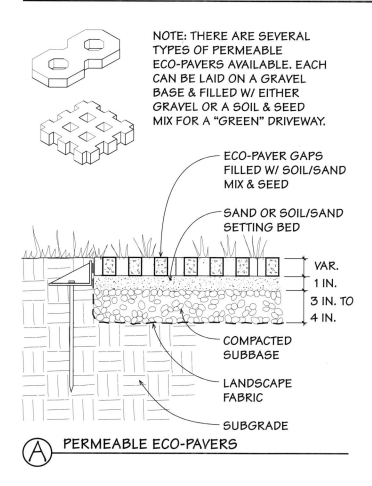

NOTE: THERE ARE SEVERAL TYPES OF PERMEABLE ECO-PAVERS AVAILABLE. EACH CAN BE LAID ON A GRAVEL BASE & FILLED W/ EITHER GRAVEL OR A SOIL & SEED MIX FOR A "GREEN" DRIVEWAY.

ECO-PAVER GAPS FILLED W/ SOIL/SAND MIX & SEED

SAND OR SOIL/SAND SETTING BED

VAR.
1 IN.
3 IN. TO 4 IN.

COMPACTED SUBBASE

LANDSCAPE FABRIC

SUBGRADE

 Ⓐ PERMEABLE ECO-PAVERS

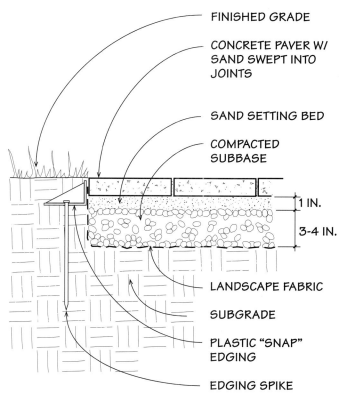

FINISHED GRADE

CONCRETE PAVER W/ SAND SWEPT INTO JOINTS

SAND SETTING BED

COMPACTED SUBBASE

1 IN.
3-4 IN.

LANDSCAPE FABRIC

SUBGRADE

PLASTIC "SNAP" EDGING

EDGING SPIKE

Ⓑ CONCRETE PAVERS ON SAND

They also offer a wide array of aesthetic possibilities, coming in many different shapes, sizes, and colors that range from browns and grays to reds and pinks. Sometimes referred to as interlocking pavers or unit pavers, many varieties are designed to create a complex pattern (see 89B). All pavers typically come in thicknesses of 2⅜ in. or 3⅜ in.; the latter is recommended for areas with vehicular traffic.

Because there are so many different paver shapes available, the number of options for unique paving patterns is nearly endless. Some patterns are designed to accommodate round or semicircular areas with no gaps between the pavers, while others are better suited to straight paths and larger, angular spaces. Like brick, pavers need a rigid edge to help hold them in place, and there are several options specifically designed for them, including one that produces a nearly invisible edge.

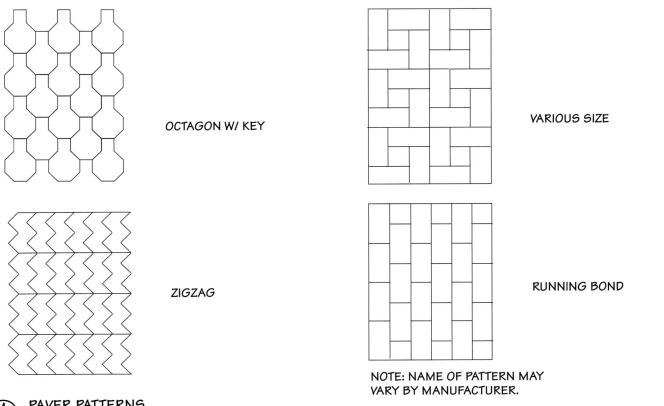

OCTAGON W/ KEY

VARIOUS SIZE

ZIGZAG

RUNNING BOND

NOTE: NAME OF PATTERN MAY
VARY BY MANUFACTURER.

Ⓐ PAVER PATTERNS

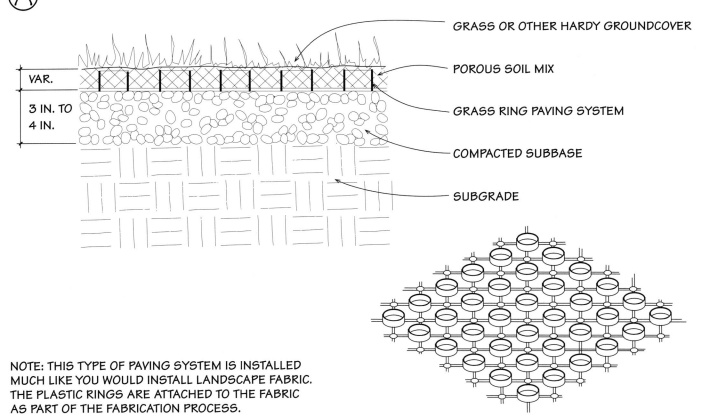

GRASS OR OTHER HARDY GROUNDCOVER

POROUS SOIL MIX

GRASS RING PAVING SYSTEM

COMPACTED SUBBASE

SUBGRADE

VAR.

3 IN. TO
4 IN.

NOTE: THIS TYPE OF PAVING SYSTEM IS INSTALLED
MUCH LIKE YOU WOULD INSTALL LANDSCAPE FABRIC.
THE PLASTIC RINGS ARE ATTACHED TO THE FABRIC
AS PART OF THE FABRICATION PROCESS.

Ⓑ GRASS RING PAVING

VEGETATIVE PAVING—A paving of sorts, vegetative-paving systems allow an area to appear as if it has been planted (most often lawn) area, even though it is capable of handling heavy traffic, such as cars and trucks. These systems work by providing large, open spaces within the hard paving material. The spaces are filled with enough sand/soil mix to support the growth of vegetation while being small enough to prevent a tire from sinking (see 90B).

Vegetative-paving systems were originally developed for use in fire lanes and other areas with infrequent traffic. Today, they're becoming popular in residential settings, usually as a substitute for all or part of a paved driveway. They are not a good material for patios or pathways, however, because the alternating concrete and groundcover creates an uneven surface.

IMPERMEABLE PAVING

Impermeable paving systems do not allow water to penetrate beneath their surface and are typically drained with sheet-flow techniques or through drainage structures (see chapter 1). Because drainage is so important, especially in the case of impermeable surfaces, care should be taken to design both the surface and the drainage system in a manner that is functional as well as aesthetically pleasing (see 91A).

ASPHALT—Typically used on driveways, roads, and recreational paths, asphalt is composed of hot tar and gravel. While asphalt is a relatively inexpensive alternative to concrete, it is much less strong and durable; a rigid paving edge should be used with it, especially in areas with vehicular traffic, to help prevent deformation and cracking.

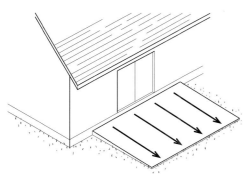

SHEET FLOW

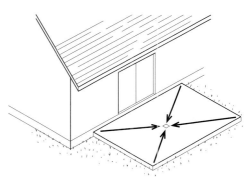

DRAINAGE TO CENTER FROM EDGES

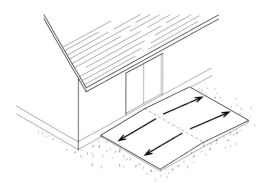

DRAINAGE FROM RIDGE IN TWO DIRECTIONS

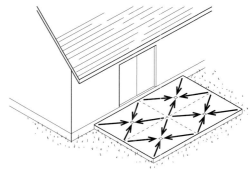

DRAINAGE AREA DIVIDED INTO SMALLER ZONES

 DRAINAGE APPROACHES FOR IMPERMEABLE PAVING SYSTEMS

Installation—Asphalt is typically installed over a compacted gravel subbase. It can be laid in a single layer or in two separate "lifts" (see 92A). The thickness and number of lifts will often depend on the type of traffic the surface will support; one, 3-in. lift is generally sufficient for a light-duty path. A 4-in. lift can be used on driveway surfaces as long as light cars are the heaviest vehicles consistently on the surface.

For driveways with heavier-duty traffic, asphalt is typically laid in two separate lifts; the first is about 2½ in. thick, and the second is about 1½ in. thick. Although significantly more costly, this method creates a stronger surface because rolling the asphalt in two separate lifts increases its compaction.

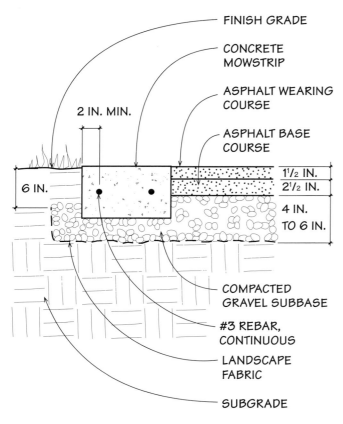

FINISH GRADE

CONCRETE MOWSTRIP

ASPHALT WEARING COURSE

ASPHALT BASE COURSE

2 IN. MIN.

6 IN.

1½ IN.
2½ IN.

4 IN.
TO 6 IN.

COMPACTED GRAVEL SUBBASE

#3 REBAR, CONTINUOUS

LANDSCAPE FABRIC

SUBGRADE

Ⓐ ASPHALT PAVING W/CONCRETE EDGE

It is not uncommon for asphalt paving to leach oil from the tar mix for several weeks after installation. Bio-bags or other oil-absorbing products should be used to prevent contamination of area waterways. Check with your local planning department or permitting agency to see if there are any restrictions.

Care and maintenance—Several things will enhance or inhibit the longevity of asphalt paving. Channeled water, such as in a gutter or paved swale, that runs over asphalt will degrade the asphalt very quickly, causing it to crumble and eventually crack and break.

Other weather-related elements, such as extreme heat or ice freeze and thaw, also can impact asphalt negatively, more so than with other paving systems. In extreme heat, asphalt can become soft and malleable, so its use in climates where temperatures routinely peak above 100°F should be considered carefully. On the opposite end of the spectrum, asphalt paving does not always hold up well in extremely cold climates either. Excessive shoveling (especially plowing) and ice-melting chemicals will quickly degrade the surface of the paving. In these conditions, asphalt paving can begin to break down within five to 10 years.

Sealing the surface of asphalt is the best thing you can do to lengthen its life. Sealant, which is generally tar based, prevents water from permeating the material and further degrading existing cracks while it also rebinds the surfaced material. Sealant should be applied every two to five years, depending on what the surface is used for and environmental conditions, such as temperature extremes.

CONCRETE—Concrete is perhaps the most versatile paving material. Its ability to take nearly any shape, color, or texture while providing an extremely durable surface has made it an attractive building material for centuries. It is relatively easy to work with, and it is very cost-effective, especially when its cost is spread across the life of the material, which can be more than 20 years.

Composition—Concrete is made up of specific proportions of crushed gravel, sand, and cement. When the three are combined with a prescribed amount of water, a chemical reaction begins that hardens the concrete. Too much of any one material can compromise concrete's strength or durability, so it is wise to let a concrete specialist help determine the right mix of material based on your job, the climate, the desired strength (measured in the pounds per square inch, or psi, that the material can support), and the nature of what you are pouring.

Most residential applications will require a 2,500-psi to 3,000-psi mix of concrete, although driveways that support heavier vehicles, such as an SUV or truck, should use a 4,000 or even 4,500-psi mix to prevent buckling or cracking. Adding reinforcing and increasing the depth of the subbase material also will strengthen the concrete (see 93A).

NOTE: USE 3,000 PSI CONCRETE FOR WALKS;
USE 3,500 TO 4,000 PSI FOR DRIVEWAYS

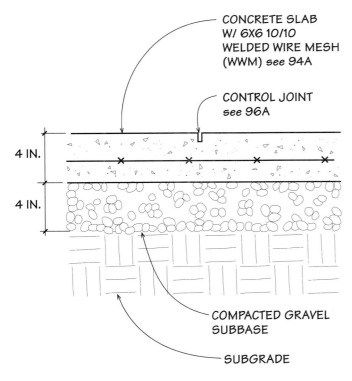

CONCRETE SLAB
W/ 6X6 10/10
WELDED WIRE MESH
(WWM) see 94A

CONTROL JOINT
see 96A

4 IN.

4 IN.

COMPACTED GRAVEL
SUBBASE

SUBGRADE

 CONCRETE PAVING

Installation—Concrete is typically poured in place with forms on top of a compacted gravel subbase. Residential concrete pours should be at least 4 in. thick; anything less is too weak and generally results in cracking and deterioration. Thicker pours can enhance the strength of the paved area, but 6 in. is generally adequate, even on surfaces with heavy vehicles. Driveways, for example, are typically poured at a depth of 4 in. to 5 in.

As concrete hardens, water evaporates from the mix, leaving behind a solid, stonelike material. Unlike stone, however, concrete will expand and contract as it hardens, and will continue doing so even after curing, in response to temperature variation. The inevitable result is cracking, which can affect functionality and aesthetics. Three critical elements, therefore, must be incorporated into the design of the concrete—reinforcing within the concrete, expansion joints, and control joints.

Reinforcing within concrete—To ensure that concrete will maintain its strength over time, even if cracking does occur, a reinforcing system of steel, fiberglass, or plastic mesh often is added to the concrete mix. Reinforcing also will help prevent differential heaving (caused by expansion and contraction during the freeze/thaw cycle).

Fibermesh reinforcing—Fibermesh is made up of small strands of fiberglass that are mixed in with the concrete at the batch plant. While not significantly important in increasing the strength of the concrete, fibermesh will help keep the concrete from pulling apart as cracks occur. Fibermesh is fine for use with many types of concrete finishes, although finishes like exposed aggregate are not recommended with fibermesh. Its most typical use is in flatwork, or level surface applications where light traffic occurs, such as on sidewalks and patios.

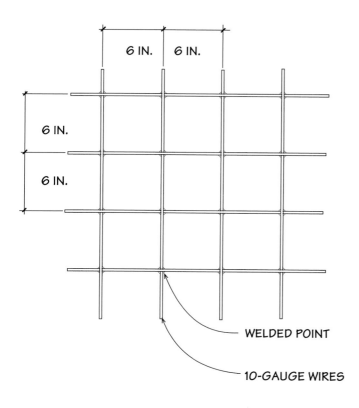

WELDED POINT

10-GAUGE WIRES

(A) WELDED WIRE MESH (WWM)

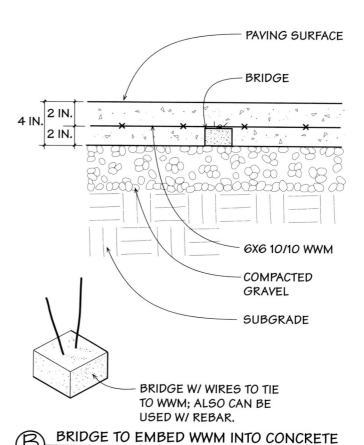

PAVING SURFACE

BRIDGE

6X6 10/10 WWM

COMPACTED GRAVEL

SUBGRADE

BRIDGE W/ WIRES TO TIE TO WWM; ALSO CAN BE USED W/ REBAR.

(B) BRIDGE TO EMBED WWM INTO CONCRETE

Welded wire mesh—Welded wire mesh (WWM) is thick, ungalvanized steel wire that is welded together in a grid pattern. It's primarily used in driveways that support average-sized cars and trucks. It is described in terms of the distance between the wires (such as 6x6, meaning wires 6 in. apart in both directions of the grid) and by the gauge of each wire (such as 10/10, meaning the wires going each way are 10-gauge wires) (see 94A).

Welded wire mesh comes in rolls and is typically cut into shapes that mirror the paving area. It is laid on top of the gravel subbase, but once pouring begins, care should be taken to make sure it sits in the middle of the concrete layer. Bridges or dobes (pronounced "doe-bees") often are used to keep the steel off of the surface of the gravel (see 94B). It is also important to keep the mesh at least 2 in. back from the edge of the concrete because moisture penetrating the edge could cause the steel to corrode, compromising its integrity.

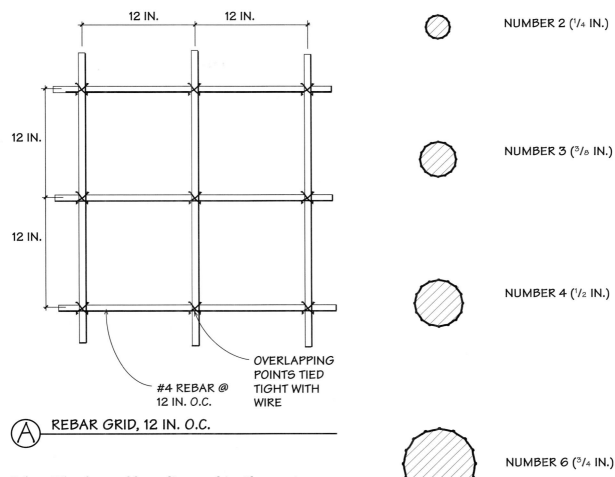

12 IN. 12 IN.

12 IN.

12 IN.

OVERLAPPING POINTS TIED TIGHT WITH WIRE

#4 REBAR @ 12 IN. O.C.

Ⓐ REBAR GRID, 12 IN. O.C.

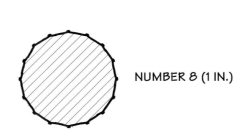

NUMBER 2 (¼ IN.)

NUMBER 3 (⅜ IN.)

NUMBER 4 (½ IN.)

NUMBER 6 (¾ IN.)

NUMBER 8 (1 IN.)

Rebar—Like the steel bars discussed in Chapter 1. Rebar, or reinforcing bars, will significantly increase the strength of the paving and enhance its ability to resist heaving over time. These steel bars, rough on the outside and nongalvanized, are embedded within the concrete much like the welded wire mesh and are typically laid out in a grid covering the area to be paved. For paving situations requiring a high degree of strength, such as a pad for motor-home parking, a grid of bars placed 12 in. on center both ways (meaning a 12x12-in. grid) is appropriate, with each intersection tied together with steel baling wire (see 95A).

Rebar is described in terms of its thickness, which is always defined in eighths of an inch. A #3 rebar, therefore, will be one with a sectional diameter of ⅜ in., while a #4 bar is ⅘, or ½ in. in diameter. The most common rebar to use in paving situations is #4, although #3 bars are often used in their place. For structural slabs a structural engineer should be consulted to determine the appropriate size and spacing of the rebar.

A. TOOLED CONTROL JOINT

B. SAWCUT CONTROL JOINT

C. EXPANSION JOINT

SMOOTH STEEL PIN @ 4 FT. O.C. MAX.

ASPHALT EXPANSION MATERIAL

NOTE: REINFORCING NOT SHOWN.

Ⓐ CONCRETE JOINTS

Control joints—Control joints are lines pressed into the surface of the concrete that direct the minor cracking that occurs during concrete's curing process. Typically, their depth is one-quarter the thickness of the concrete, and they are either troweled in with a special tool while the concrete is still wet or cut in with a saw within 24 hours after the concrete has dried (see 96A). Sidewalks and other linear pathways should have a control joint every 5 ft., while larger paved areas should have them no farther than 8 ft. to 10 ft. apart.

Control joints can be a wonderful design element as well. Taking design cues from surrounding architectural details or features, you can use control joints to create beautiful visual patterns on the ground that simultaneously serve an aesthetic and functional role.

Expansion joints—Expansion joints also can serve an aesthetic as well as functional role, but they are even more critical than control joints. Expansion joints provide a zone of flexibility in the concrete that prevents buckling and cracking when expansion and contraction do occur. Typically made of a fiberboard impregnated with asphalt, these joints sit in between

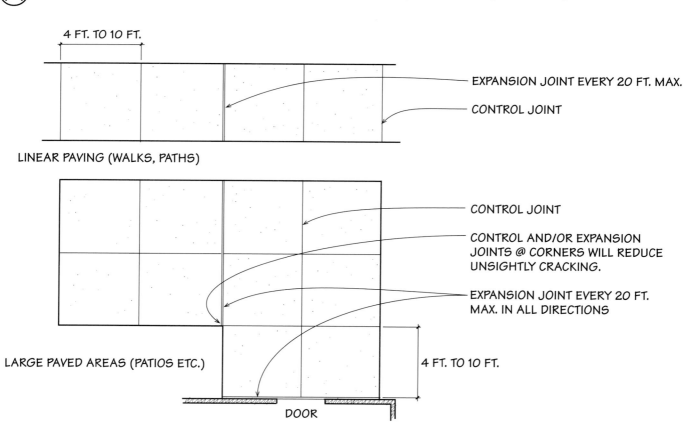

4 FT. TO 10 FT.

EXPANSION JOINT EVERY 20 FT. MAX.

CONTROL JOINT

LINEAR PAVING (WALKS, PATHS)

CONTROL JOINT

CONTROL AND/OR EXPANSION JOINTS @ CORNERS WILL REDUCE UNSIGHTLY CRACKING.

EXPANSION JOINT EVERY 20 FT. MAX. IN ALL DIRECTIONS

LARGE PAVED AREAS (PATIOS ETC.)

4 FT. TO 10 FT.

DOOR

Ⓑ CONTROL JOINTS AND EXPANSION JOINTS

slabs of concrete and prevent them from pressing on one another. They should be installed every 20 ft. along a pathway, and no paved area should have a length greater than 20 ft. without an expansion joint (see 96B). Expansion joints also must be used where a concrete slab abuts a solid object such as a building or a preexisting slab or wall.

Finishes—Once the concrete is in place, there are a number of choices available for finishing the surface of the concrete, including colors, textures, and patterns.

Colors—Coloring concrete has become very popular over the past 10 years, and there are now many choices in terms of both color and application technique.

Concrete stain, much like furniture stain, can be applied to the surface of recently poured concrete. It stains the top $\frac{1}{16}$ in. to $\frac{1}{8}$ in. of material, creating an almost translucent appearance. It is difficult to apply and toxic in nature, requiring ventilation equipment and protective suits and gloves. All runoff of chemicals should also be carefully controlled, collected, and properly disposed of. Also, because only the surface is colored, stain is not recommended in high-traffic areas that could be chipped or scratched.

Shake-on colors are a powdery substance applied to wet concrete and troweled into the surface. Like stain, only the top layers (in this case, $\frac{1}{8}$ in. to $\frac{1}{4}$ in.) of concrete are colored, making chips a potential problem. It is less costly than integral color (see below), however, and it comes in a wider array of colors, including many darker hues. A general rule is the darker the hue, the more expensive the concrete per square ft.

Integral colors are also a powdery substance, but they are mixed into the concrete by a concrete supplier. Since the color is mixed throughout the slab, chipping is far less problematic, variation in color is minimized, and application requires no extra step. It is, however, more costly, and there is a smaller palette of colors to choose from.

Stamped patterns—Often combined with integral coloring systems, stamped concrete offers a wide variety of patterns that generally mimic other, more expensive paving patterns at a fraction of the cost.

Once the concrete is poured in place, a series of stamps are pressed into the still-wet concrete. The patterns range from brick to flagstone to cut-stone laid in a fish-scale pattern. The patterns generally eliminate the need for control joints, but expansion joints are still necessary.

Textures—Before there were color and pattern options for concrete, texturing the surface was the primary method used for finishing concrete. While there are many variations, the most prominent are described here.

• **Broom Finish**—Created by lightly dragging a soft-bristled broom over slightly wet concrete, broom finishes are slip resistant and can produce a very pleasing, subtle aesthetic. Typically, the broom is dragged consistently in one direction perpendicular to the main circulation route.

• **Acid-etched**—Once the concrete has hardened, the surface is washed, or etched, with a mixture of water and hydrochloric acid. Scrubbed in hard with a stiff bristle broom, this method washes away the top surface "cream" of the concrete and exposes just the top pieces of the aggregate. The acid used in etching is highly toxic and can result in significant injury if proper gear isn't worn (rubber boots, gloves, long pants, long sleeves, goggles). Surrounding plants also should be protected.

• **Sand-blast finish**—This method uses sand blown at extremely high pressure to erode the top surface of the concrete, creating a rough, almost jagged finish. This finish can be used to create patterns by covering some areas and sandblasting others, much like stenciling. Done carefully, it can produce a very attractive finish, but overdoing it creates a very rough and uncomfortable surface.

• **Troweled finish**—This method is fine for very small areas, but it is typically difficult to make attractive or slip resistant. Trowel finishes often are used as an intermediate step before another finish, such as broom or acid-etch, is applied.

BROOM

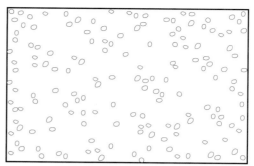

EXPOSED AGGREGATE

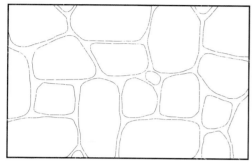

STAMPED PATTERN

ACID-ETCHED

 CONCRETE FINISHES

• **Exposed aggregate**—Exposed aggregate finishes use the gravel within a concrete mix to create a textured surface. Once the concrete is poured, it is troweled as little as possible so the gravel remains close to the surface. Once the concrete begins to set, it is sprayed with water to remove the cream, exposing the gravel. Care should be taken to not expose more than the top one-quarter to one-third of the rock to prevent dislodging, or "blowout." Specific gravels can be used in a concrete mix to create colored aggregate finishes, but this is more costly.

A variation of this finish, called "seeding," involves sprinkling a gravel mix over still-wet concrete and lightly pressing it into the surface. This is a much cheaper way to achieve a particular color aggregate finish, but it is much more susceptible to blowout.

TILE, BRICK, AND STONE OVER CONCRETE—Several of the paving materials described in the permeable section of this chapter also can be installed over an impermeable concrete subbase. Because this tends to be such a hands-on, time-intensive process, it is typically the most expensive category of paving options.

To achieve this type of paving, a concrete subbase is first poured in place over a gravel layer. With tile, a thin grout layer is then troweled over the top, and the tiles are positioned. Once the tile and grout have set, a fill-material grout (available in a wide range of colors) is then pressed into the gaps between the tiles; the excess is wiped clean.

Bricks, stone, and even some tile materials are set into a mortar bed (½ in. to ¾ in. in depth) instead of grout (see 99A). Stones and brick should be damp (not soaked) prior to installation so they don't absorb the moisture from the mortar setting bed too quickly, weakening the connection. Once the stone or brick has set, mortar can be troweled or "mortar-bagged" (similar to an icing bag) into the gaps. The mortar is then tooled into place to remove any excess and ensure that the mortar is completely packed into the spaces between the units.

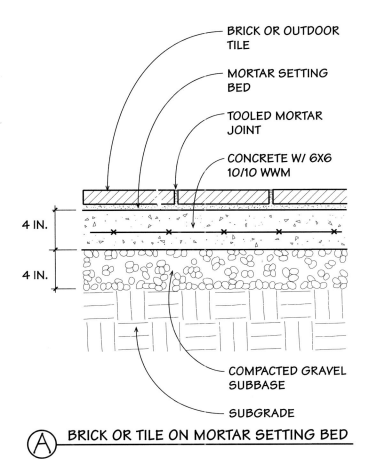

BRICK OR OUTDOOR TILE

MORTAR SETTING BED

TOOLED MORTAR JOINT

CONCRETE W/ 6X6 10/10 WWM

4 IN.

4 IN.

COMPACTED GRAVEL SUBBASE

SUBGRADE

(A) BRICK OR TILE ON MORTAR SETTING BED

CHAPTER 4

LANDSCAPE STRUCTURES

Outdoor structures that are constructed upon the ground—the decks, fences, arbors, and gazebos that make landscaped spaces more useful and beautiful—are diverse but have two characteristics in common. First, they are built to stand on their own above the ground, so their structural capacity must be carefully considered. Second, they are subjected to the harsh effects of the weather during all seasons, so their materials must be carefully selected and their joints detailed to minimize deterioration.

PRINCIPLES OF STRUCTURE

There are three basic types of forces that impact structures: vertical loads (gravity), lateral loads (wind and earthquakes), and uplift loads (primarily wind). Structures must be designed to resist these in order to remain stable.

VERTICAL LOADS—Structures built upon the landscape need to be adequately supported by the ground so that they do not sink into it. This is typically accomplished by spreading the weight of the structure over a large area of soil. Most commonly, columns from the structure are connected to and supported by a concrete spread footing.

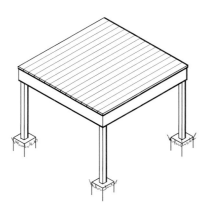

Sometimes the structural loads are minimal and can be supported on small, precast concrete piers. In other cases, the magnitude of the loads is great, and large, cast-in-place concrete footings are required.

The required size of a footing depends on the load that it supports and the type of soil on which it sits. With general guidelines for both of these variables, footing size can be easily approximated. For example, a typical deck structure weighs a maximum of 10 lb. per square foot (psf), and people and furniture on the deck weigh a maximum of 40 psf. Together, the deck and the load it carries weigh a total of 50 psf. The weight of a fully loaded 10x12-ft. area of deck, then, would impose a load of about 6,000 lb. (120 sq. ft. x 50 psf) on the ground.

Most soils will support 1,500 psf to 2,000 psf. Thus, being conservative, we would need at least 4 sq. ft. of footing (6,000 lb./1,500 psf) to support a 10x12-ft. area of deck on typical soil. If this area constituted the entire deck and was supported at all four corners, then each of the four footings would only need to be 1 sq. ft.—about the size of a common precast pier.

LATERAL LOADS—Structures built upon the landscape also need to be stabilized so that they do not fall over sideways. There are three methods commonly employed to accomplish this. First, diagonal braces

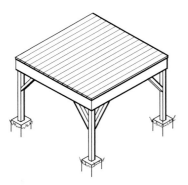

DIAGONAL BRACING

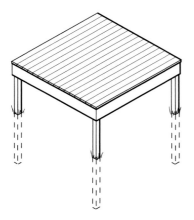

EMBEDDED COLUMNS

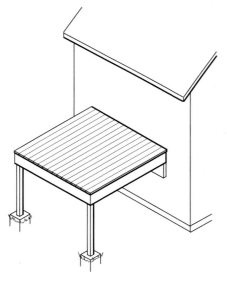

ATTACHMENT TO STABLE ELEMENT

can be used to stabilize the columns of a structure. Second, columns can be embedded into the ground like the roots of trees to stabilize them and the structure they support. Finally, structures can be attached to stable elements such as buildings to make them stable themselves. The design of the structure can utilize two or more methods to maximize stability.

UPLIFT LOADS—Landscape structures also need to be designed to resist uplift loads—primarily the result of wind getting under the roof of a structure and pushing it up. In regions particularly known for high winds, structures should be firmly anchored to the ground with a heavy foundation or one that is firmly rooted in the ground. Furthermore, connections of all structural members should be designed with bolts, straps, or interlocking pieces to resist vertical loads in the upward direction.

PRINCIPLES OF WEATHERING

Most landscape structures are exposed to the degenerating effects of the weather (moisture, sun, wind, and salts) and will not last as long as roofed structures. So it is important to design and detail them carefully to minimize decay and extend their useful life. Primary concerns in this regard are material selection and detailing—especially of joints.

MATERIAL SELECTION—Materials that are only minimally affected by moisture are most durable for landscape structures. Masonry, concrete, synthetic wood, and wrought iron are best. Steel with a proper finish coating will also last a very long time. All these materials are also unaffected by insect pests such as termites, a critical consideration in warm, humid areas.

Wood, however, because of its availability, workability, beauty, and relatively low cost, remains a favorite for outdoor structures. But wood absorbs moisture readily, which causes it to expand, twist, and check, increasing its susceptibility to rotting.

Natural wood—Naturally weather-resistant wood species, such as cedar, redwood, cypress, teak, and locust, are appropriate for outdoor use because they will not rot as rapidly or readily as other wood species.

Preservative-treated wood—Preservative-treated (pressure-treated) wood will outlast most naturally resistant species but in time will also deteriorate, because the factory-injected treatment does not usually penetrate fully to the core of the wood. Field-applied preservative treatments are also available but are not as effective as those applied at the factory. Applying a transparent stain to preservative-treated wood can increase its durability by a factor of two to three.

Factory-applied treatments are identified with a stamp on each piece of lumber. Among other things, the stamp describes the intended exposure condition (above ground or ground contact), the date of treatment, and the preservative chemical. The most widely used chemical is chromated copper arsenate (CCA), easily identified by its green hue. Other common treatments include ammoniacal copper arsenate (ACA) and ammoniacal copper zinc arsenate (ACZA). These three treatments are the most toxic and can leach into the ground.

Ammoniacal quaternary ammonia (ACQ) and ammoniacal copper citrate (ACC) are newer treatments that contain no hazardous chromium or arsenic. Lumber treated with these chemicals is slightly more expensive than lumber treated with the more widely used chemicals.

Synthetic wood—Made of reclaimed hardwood fibers and recycled plastic, synthetic wood is very durable and stable in exposed conditions. It is not harmed by rot or insects and does not absorb water. It is not as strong as natural lumber, so its structural use is limited, but it works well for decking.

CONNECTION DESIGN—The points of connection between wood members are susceptible to moisture buildup, making the wood especially prone to rot. Even preservative-treated lumber will rot in this loca-

tion. Therefore, to minimize rot in exposed conditions, it is advisable to follow some basic design principles:

- Avoid doubling up members in exposed situations. It is better to use a single large timber where extra strength is required.

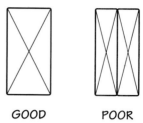

GOOD   POOR

- Where wood must touch another surface, make the area of contact as small as possible and allow for air circulation around the joint.

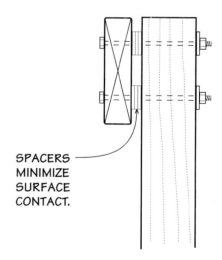

SPACERS MINIMIZE SURFACE CONTACT.

- Attempt to avoid wide horizontal surfaces because they will hold moisture from rain or snow and lead to rapid deterioration of the wood. Where possible, make horizontal surfaces narrow or slope them to drain.

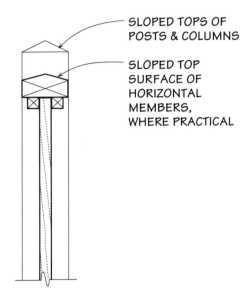

SLOPED TOPS OF
POSTS & COLUMNS

SLOPED TOP
SURFACE OF
HORIZONTAL
MEMBERS,
WHERE PRACTICAL

FASTENERS—At least once a year, joints that are exposed to the weather will shrink and swell, causing nails to withdraw and joints to weaken. Joints fastened with screws or bolts will outlast those connected by nails, but the shrinking and swelling of the joint will crush wood fibers and reduce their strength, causing looseness in the joint. Screwed and bolted joints can be tightened periodically, restoring them to their approximate original strength. Withdrawn nails can be redriven, but the connection will never be as strong as it was originally.

Metal fasteners such as nails, bolts, and screws should be galvanized to prevent rust; hot-dip-galvanized nails resist withdrawal much better than electro-galvanized nails. Stainless-steel and brass screws provide the longest life, but they're also more expensive than galvanized fasteners.

Galvanized steel connectors such as joist hangers, post bases, and angle clips are especially useful for making strong connections. These connectors are designed for strength rather than beauty, however, so their use is typically restricted to locations where they are hidden from view—below decks and porches, for instance.

Galvanized steel deteriorates rather quickly when combined with preservative-treated lumber, especially when moisture is added to the mix, so an annual inspection of galvanized connectors is recommended where safety issues are a concern.

HORIZONTAL BOARDS—Horizontally oriented boards, such as decking, have a relatively large exposed surface to collect and absorb moisture. This moisture will tend to make the boards cup. Flat-grain boards should be placed with the bark side up so if they cup, they will shed water.

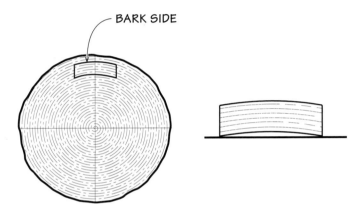

BARK SIDE

The movement of decking boards can be minimized by limiting their width to 6 in. (using 2x6 lumber) and by fastening them with deck screws or ring-shank nails, both of which have very high withdrawal resistance.

COATINGS—Sealers and preservatives will extend the life of exposed landscape structures. Extra coats should be applied to the end grain of wood and to other areas likely to absorb and hold moisture.

Penetrating stains will outlast paints because they are absorbed into the material and do not build up on the surface. Though paints offer more brilliant color than stains, they form a surface film that can trap moisture when the paint starts to deteriorate.

Coatings with a water-resistant component such as silicone will extend the life of exposed wood, concrete, or masonry by limiting the entry of moisture beyond the surface of the material. This will minimize the opportunities for decay-producing organisms and can also virtually eliminate damage due to freezing temperatures. To be effective, however, water-resistant coatings generally must be reapplied every few years.

WOOD DECKS

A wood deck is a common and practical device for extending the living area from inside the house to the outdoors. Although not as durable as a ground-level terrace made of concrete or pavers, a wood deck can be made to last for decades when designed carefully. A wood deck is especially economical in comparison to a ground-level terrace in locations where the ground slopes significantly.

The design of a deck involves myriad decisions. To be truly useful, it must be located appropriately in relation to the house, the yard, long-distance views, the sun, and the prevailing wind. Its size and shape, also important considerations in relation to its usefulness and beauty, are typically and logically influenced by the 2-ft. increments of standard lumber. Other significant design considerations include:

- the structure of the deck: How will the structural system work? How will this be supported by the ground, and how will the deck be laterally braced?

- building and/or zoning codes: How will they apply to the design?

- the material used for the deck surface: What grade of wood or kind of synthetic decking will be used? How will it be fastened to the framing?

- the appearance of the deck from the yard: Will the underside of the deck be visible or will a skirt between deck and ground hide the structure?

- deck maintenance: Will access to the underside of the deck be necessary? Will there be a need to control weed growth under the deck? Will preservatives and sealers be required for long-term durability?

- the design of stairs and guardrails: If required, how will these elements be supported, and how will they complement the aesthetic of the deck itself?

The resolution of these issues is rarely simple, but as much detail as possible should be determined before construction begins, because a decision about one issue will inevitably affect other issues.

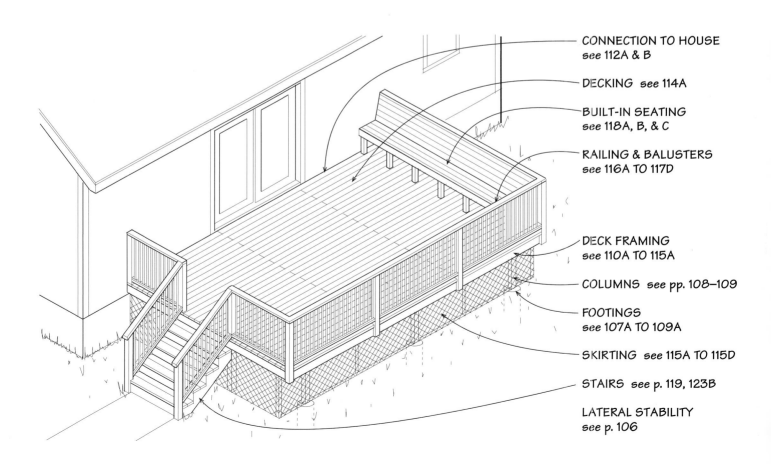

CONNECTION TO HOUSE
see 112A & B

DECKING see 114A

BUILT-IN SEATING
see 118A, B, & C

RAILING & BALUSTERS
see 116A TO 117D

DECK FRAMING
see 110A TO 115A

COLUMNS see pp. 108–109

FOOTINGS
see 107A TO 109A

SKIRTING see 115A TO 115D

STAIRS see p. 119, 123B

LATERAL STABILITY
see p. 106

LATERAL STABILITY—Lateral support prevents a deck from collapsing sideways. Adequate lateral support can be achieved by stabilizing (any) two edges of a rectangular deck. Thus, if a deck is attached to a house on two of its four sides, there is no need for further lateral support. For a freestanding deck with no edge connected to a house, lateral support is usually achieved by means of diagonal bracing. In this case, each corner must be braced or attached to another corner so that it cannot move in any direction. With only one edge of a deck stabilized by the house, the two corners away from the house must be braced.

FREESTANDING DECK

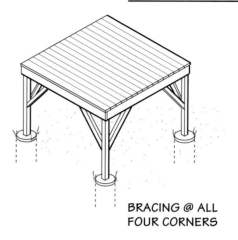

BRACING @ ALL
FOUR CORNERS

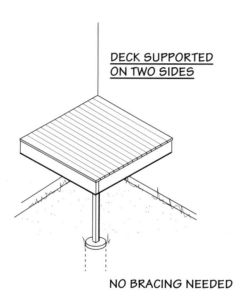

DECK SUPPORTED
ON TWO SIDES

NO BRACING NEEDED

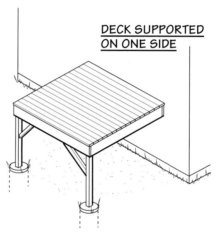

DECK SUPPORTED
ON ONE SIDE

BRACING IN ONE DIRECTION
@ OUTSIDE CORNERS

The diagonal bracing of small decks that are relatively close to the ground is often done intuitively. Large diagonal braces for decks high off the ground should be engineered in accordance with the life safety policies outlined in the building codes.

FOUNDATIONS AND FOOTINGS—Foundation support for decks is typically achieved with independent footings called column footings (or pier pads) at the base of each column or post. The primary purpose of a foundation system is to provide vertical support for the deck—keeping it level, minimizing settlement, and preventing uplift from frost, expansive soils, or wind. All footings should be placed on unfrozen, undisturbed soil that's free of organic material. The bottom of the footing must be located below the frost line. Frost lines range from 0 ft. to 6 ft. in the continental United States. Check your local building department for frost-line requirements (see 107A).

Determining the size of footings always involves guesswork, so it is prudent to err on the side of caution. The size of the footing depends on two basic factors:

• the assumed load of the deck—its self-weight (dead load) and the weight of people and furniture upon it (live load);

• the load-bearing capacity of the soil. Soils are tested and rated as to their ability to support loads.

Soil Type	Bearing Capacity (psf)
Soft Clay or Silt	Do Not Build
Medium Clay or Silt	1,500-2,200
Stiff Clay or Silt	2,200-2,500
Loose Sand	1,500-2,000
Dense Sand	2,000-3,000
Gravel	2,500-3,000
Bedrock	4,000 and up

Compaction of soil may be required before footings are placed. The addition of compacted rock under a footing will also increase its bearing capacity. Consult with the house's foundation designer for the bearing capacity of the soil.

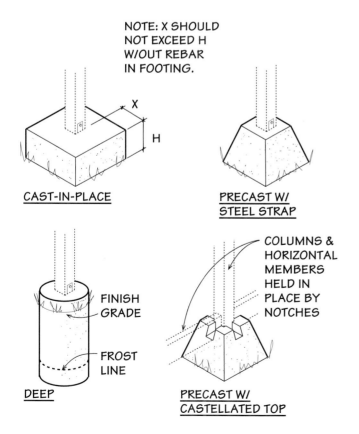

NOTE: X SHOULD NOT EXCEED H W/OUT REBAR IN FOOTING.

CAST-IN-PLACE

PRECAST W/ STEEL STRAP

DEEP

FINISH GRADE

FROST LINE

COLUMNS & HORIZONTAL MEMBERS HELD IN PLACE BY NOTCHES

PRECAST W/ CASTELLATED TOP

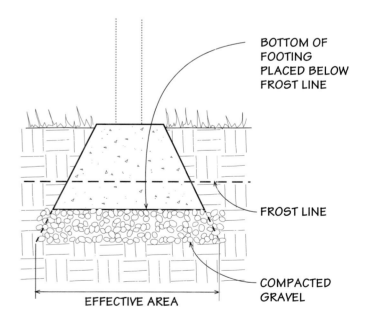

BOTTOM OF FOOTING PLACED BELOW FROST LINE

FROST LINE

COMPACTED GRAVEL

EFFECTIVE AREA

 COLUMN FOOTINGS

A conservative estimate is that the deck and its load of people and furniture will weigh no more than 50 pounds per square foot (psf). Given that the weakest soil suitable for building upon will support 1,500 psf, and a fully loaded deck will weigh only 50 psf, the assumption can be made that 1 sq. ft. of footing will support 30 sq. ft. of deck (1,500 psf/50 psf =30 sq. ft.).

For decks that are reasonably close to ground level, it is practical to use a large number of small, closely spaced column footings to support the deck. For decks that are high above the ground, it is sensible to consolidate the vertical forces into a few columns placed on large column footings.

COLUMNS—Columns (or posts) bear on the center of column footings and extend up vertically to a level plane upon which the deck is built. In the simplest case, a 4x4 column rises only 1 ft. or so to support a beam under the deck (see 108A). This simple column is nailed to the beam and rests on the column footing where it is bolted or nailed to a metal strap. When columns are tall or need to resist the uplift forces of high winds or the lateral forces of earthquakes, they are more securely fastened at their base to the column footing and at their top to the beam they support (see 109A & B). Tall columns also need to be engineered to determine that they will not buckle under the weight of the deck, and they may need diagonal bracing for lateral support.

Because of their important structural role and proximity at their base to soil and vegetation, deck columns are usually made of preservative-treated wood to ensure they don't rot.

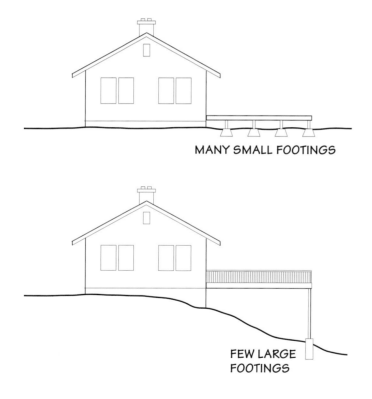

MANY SMALL FOOTINGS

FEW LARGE FOOTINGS

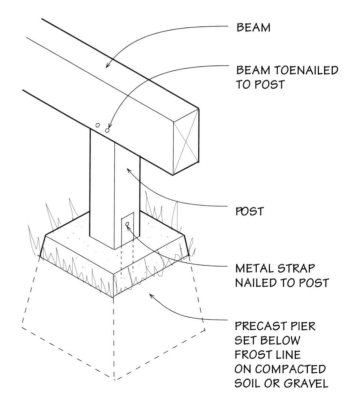

BEAM

BEAM TOENAILED TO POST

POST

METAL STRAP NAILED TO POST

PRECAST PIER SET BELOW FROST LINE ON COMPACTED SOIL OR GRAVEL

Ⓐ **BEAM/FOOTING CONNECTION**
SHORT POST ON PRECAST PIER

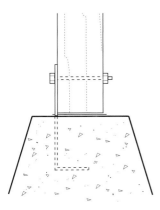

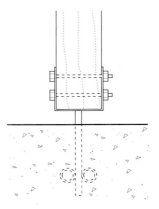

SINGLE STRAP

GALVANIZED STEEL
STRAP CONNECTOR IS
OFTEN FOUND ON
PRECAST PIER BLOCKS.
BEST USED FOR SHORT
COLUMNS. BOLT STRAP
TO COLUMN & LOCATE
MOISTURE BARRIER
BETWEEN COLUMN
& FOOTING.

CAST-IN-PLACE

A VARIETY OF STEEL
COLUMN BASES ARE
AVAILABLE. THEY ARE
SET INTO CONCRETE
FOOTINGS, ISOLATING
THE COLUMN FROM THE
CONCRETE & PROVIDING
A RIGID CONNECTION.
BOLT COLUMN TO BASE.

 COLUMN BASES

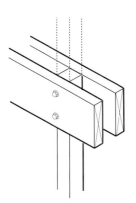

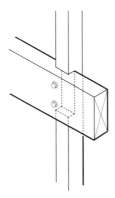

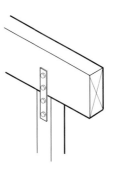

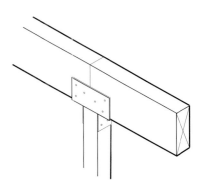

BEAM BOLTED TO SIDE OF COLUMN

USING BOLTS ON
EITHER SIDE OF
COLUMN ALLOWS
FOR THE USE OF
SMALLER- DIMENSION
LUMBER. THIS
TECHNIQUE ALSO
ALLOWS COLUMN TO
CONTINUE THROUGH
FOR USE IN RAIL
CONSTRUCTION.

BEAM LET INTO NOTCHED COLUMN

NOTCHING COLUMN
PROVIDES A SHELF
FOR BEAM TO SIT ON.
BOLTING CONNECTION
W/ BOLTS PROVIDES
A VERY RIGID JOINT.
THIS TECHNIQUE ALSO
ALLOWS COLUMN TO
CONTINUE THROUGH
FOR USE IN RAIL
CONSTRUCTION.

BEAM STRAPPED TO TOP OF COLUMN

AS SHOWN IN 108A,
A BEAM CAN BE
SIMPLY TOENAILED
INTO A POST OR
COLUMN. HOWEVER,
FOR TALLER DECKS IT
IS A GOOD IDEA TO
BOLT TOGETHER W/
STEEL STRAP
CONNECTORS ON
@ LEAST ONE SIDE.

BEAMS SPLICED ON COLUMN

SPLICED BEAMS
SHOULD REST DIRECTLY
ON COLUMN & BE
CONNECTED TO EACH
OTHER & THE COLUMN
W/ ONE OF A VARIETY
OF METAL CONNECTORS
THAT ARE ATTACHED
W/ NAILS OR SCREWS.

 BEAM/COLUMN CONNECTIONS

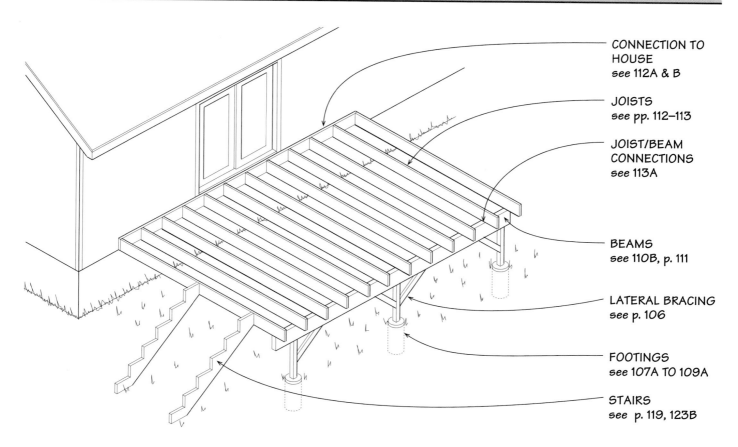

CONNECTION TO HOUSE
see 112A & B

JOISTS
see pp. 112–113

JOIST/BEAM CONNECTIONS
see 113A

BEAMS
see 110B, p. 111

LATERAL BRACING
see p. 106

FOOTINGS
see 107A TO 109A

STAIRS
see p. 119, 123B

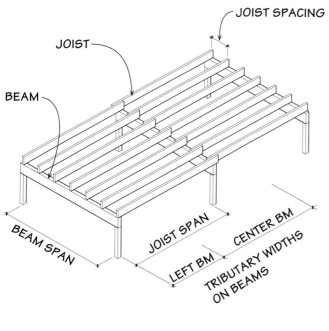

JOIST SPACING

JOIST

BEAM

BEAM SPAN

JOIST SPAN

LEFT BM

CENTER BM

TRIBUTARY WIDTHS ON BEAMS

FRAMEWORK—The horizontal structure of decks is commonly made of primary supports (beams) and secondary supports (joists). In the simplest terms, the beams are sized to carry the joists, and the joists are sized to carry the decking. Beams, joists, and decking are all sized to carry the loads of people, furniture, and snow.

Beams—Most residential deck beams are made of 4x material—4x6 through 4x12. The size of the beam is determined by the species of wood, the load from the joists, and the span of the beam between supports. As the span of the joists increases, so does the load from the joists, and the beam must be accordingly larger. Likewise, as the beam span increases, it must be larger (see the table on facing page).

Deck Beam Maximum Spans (Ft.)				
Beam Size and Type	Tributary Width			
	6 ft.	8 ft.	12 ft.	16 ft.
4x6 P.T. Timber	6.1	4.6	3.0	2.5
4x8 P.T. Timber	8.0	6.0	4.0	3.0
4x10 P.T. Timber	10.2	7.6	5.1	3.8
4x12 P.T. Timber	12.1	9.3	6.2	4.7
4x10 P.T. Parallam	13.6	12.5	10.6	8.8
4x12 P.T. Parallam	17.0	15.4	13.0	11.0

NOTE: Beam spans are calculated for wet use in typical residential loads and are for estimating purposes only.

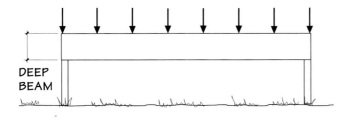

DEEP BEAM

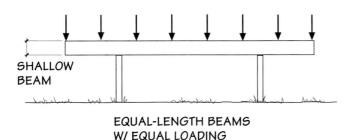

SHALLOW BEAM

EQUAL-LENGTH BEAMS
W/ EQUAL LOADING

Deck beams are typically supported on columns (see 109B). The strongest connections provide bearing for the beam directly over the column. The beam also may be bolted to the side of a column, but this arrangement is not as strong in the long run because the bolts (and the relatively small area of wood fiber around the bolts) must support the entire load of the beam. In a hybrid connection, the beam is fitted into a notch in the column, partially bearing on it as well as being bolted in.

Because of the flexibility in locating vertical supports for decks, a unique opportunity exists to take advantage of the structural efficiency of the cantilever—a structure that projects horizontally beyond its support. For example, a deck that is 16 ft. wide can employ a beam that is much smaller if the columns are located at the one-quarter point rather than at the ends of the beam (see the diagram at left). This same principle holds true for joists, which can be designed to cantilever beyond supporting beams.

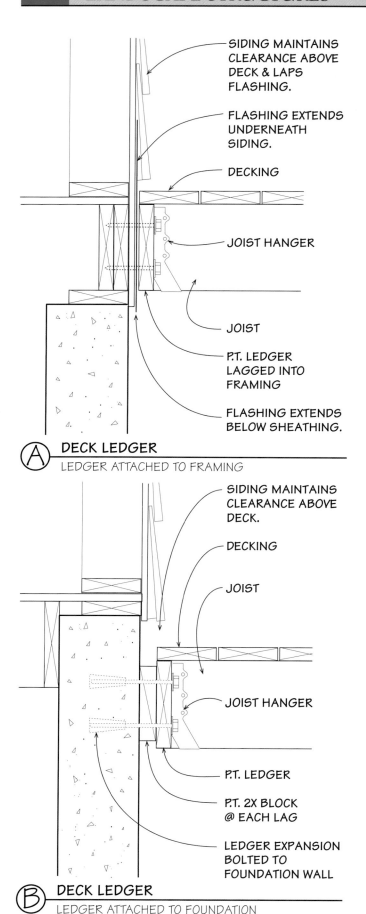

SIDING MAINTAINS CLEARANCE ABOVE DECK & LAPS FLASHING.

FLASHING EXTENDS UNDERNEATH SIDING.

DECKING

JOIST HANGER

JOIST

P.T. LEDGER LAGGED INTO FRAMING

FLASHING EXTENDS BELOW SHEATHING.

Ⓐ **DECK LEDGER**
LEDGER ATTACHED TO FRAMING

SIDING MAINTAINS CLEARANCE ABOVE DECK.

DECKING

JOIST

JOIST HANGER

P.T. LEDGER

P.T. 2X BLOCK @ EACH LAG

LEDGER EXPANSION BOLTED TO FOUNDATION WALL

Ⓑ **DECK LEDGER**
LEDGER ATTACHED TO FOUNDATION

Ledgers—One or more edges of a deck are often supported at the house by a ledger (see 112A & B). Because a ledger can be fastened at frequent intervals, it generally doesn't have to be as large as the beams that span from column to column. Ledgers also provide lateral as well as vertical support (see p. 106).

Joists—Most residential deck joists are made of 2x material—2x6 or 2x8, and occasionally deeper. The size of the joist is determined by the species of wood, the spacing between joists, and the span of the joists between supports. As the spacing of the joists increases, so does the load from the decking, and the joists must be accordingly deeper. Likewise, as the span of joist increases, they must also be relatively deeper.

Deck Joist Maximum Spans (Ft.)			
Joist Size and Type	**Joist Spacing**		
	12" O.C.	**16" O.C.**	**24" O.C.**
2x6 P.T. Hem-Fir #2	9.6	8.7	7.6
2x6 Cedar #2	8.8	8.0	7.0
2x8 P.T. Hem-Fir #2	12.7	11.5	10.1
2x8 Cedar #2	11.6	10.6	9.2
2x10 P.T. Hem-Fir #2	16.2	14.7	12.8
2x10 Cedar #2	14.8	13.5	11.8
2x10 P.T. Parallam	19.6	17.8	15.6
2x12 P.T. Hem-Fir #2	19.7	17.9	15.1
2x12 Cedar #2	18.0	16.4	13.7
2x12 P.T. Parallam	24.5	22.3	19.5

NOTE: Joist spans are calculated for wet use in typical residential loads and are for estimating purposes only.

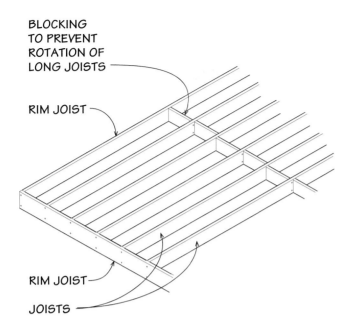

BLOCKING TO PREVENT ROTATION OF LONG JOISTS

RIM JOIST

RIM JOIST

JOISTS

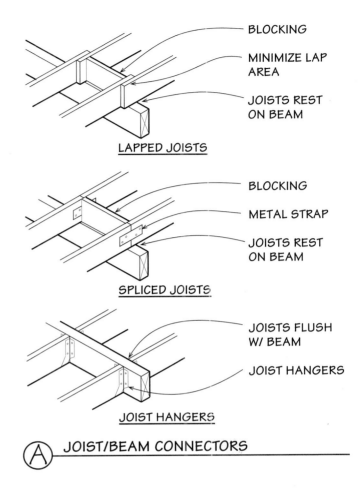

BLOCKING

MINIMIZE LAP AREA

JOISTS REST ON BEAM

LAPPED JOISTS

BLOCKING

METAL STRAP

JOISTS REST ON BEAM

SPLICED JOISTS

JOISTS FLUSH W/ BEAM

JOIST HANGERS

JOIST HANGERS

Ⓐ JOIST/BEAM CONNECTORS

The open ends of joists are typically covered with a rim joist. This perpendicular member serves to hold the joists in place and prevent them from rotating while also providing a clean appearance for the edge of the deck. At the center of decks, joists may be supported on beams in a variety of ways (see 113A).

Openings in decks for trees or other landscape features (unless very small) must be framed with structure on all sides. Joists that are interrupted to make the opening are supported on headers that, in turn, are supported by parallel joists adjacent to the opening. Where the opening is large and loads are great, the size of framing members will have to be increased.

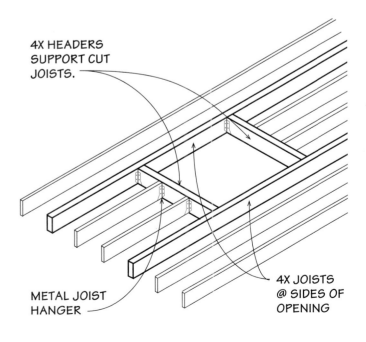

4X HEADERS SUPPORT CUT JOISTS.

METAL JOIST HANGER

4X JOISTS @ SIDES OF OPENING

Ⓑ FRAMING FOR OPENING IN DECK

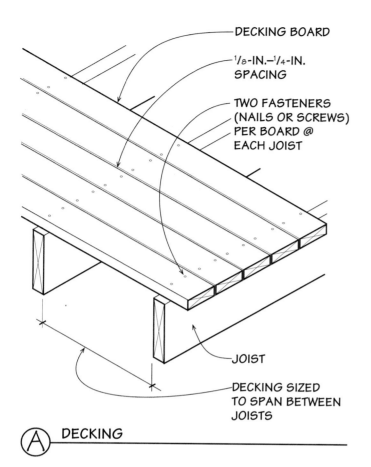

DECKING BOARD

$^1/_8$-IN.–$^1/_4$-IN. SPACING

TWO FASTENERS (NAILS OR SCREWS) PER BOARD @ EACH JOIST

JOIST

DECKING SIZED TO SPAN BETWEEN JOISTS

Ⓐ DECKING

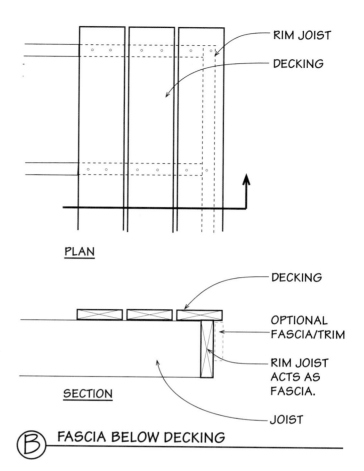

RIM JOIST

DECKING

PLAN

DECKING

OPTIONAL FASCIA/TRIM

RIM JOIST ACTS AS FASCIA.

JOIST

SECTION

Ⓑ FASCIA BELOW DECKING

Decking—The boards that form the surface of a deck are called decking. Because decking is oriented horizontally, it has a relatively large exposed surface to collect and absorb moisture. This moisture will tend to make the decking cup, especially if the decking is wide. Therefore, wooden decking is generally limited to 2x4 or 2x6 widths. Typically, decking will span between joists spaced at 24 in. on center. Individual decking boards are spaced approximately ¼ in. from each other to allow for expansion and for water to pass through the deck (see 114A).

Synthetic decking, available in standard sizes from 1x6 to 2x8, does not absorb moisture, so cupping is not a concern. It requires no sealers or preservatives and is manufactured with a nonskid surface. The material is not as strong as natural lumber of the same dimension, so joists often must be spaced closer together. And thermal expansion is a concern, so gaps must be left between the ends of boards to allow for the expansion of long runs of synthetic decking.

Fasteners—Hot-dipped galvanized nails work well to fasten wooden decking to joists, but deck screws are better because they will not withdraw due to expansion and contraction of the decking. Synthetic decking always should be fastened with deck screws.

Deck edge—The edge of a deck is frequently exposed as a part of the aesthetic of the landscape, in which case its detailing is an important thing to consider. A deck edge can be finished simply with the joists and decking that are the major elements of its construction, or other elements can be added to make it more refined (see 114B & 115A).

Skirting—The space beneath a deck is usually concealed with skirting (see 115B to D). Attached at its top to the edge of the deck and at its base to a preservative-treated framing near the ground, skirting provides a visual screen while allowing air to circulate under the deck. Skirting often is made of lattice, but also can be made of structural panels or of simple boards. When skirting is used, access to the underside of the deck should be maintained to remove debris and unwanted plants.

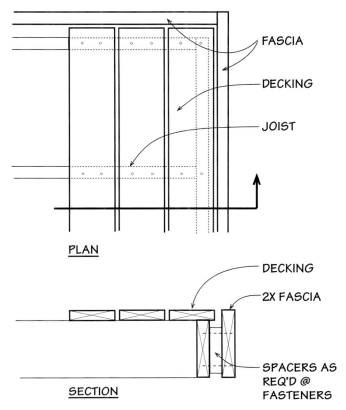

FASCIA

DECKING

JOIST

PLAN

DECKING

2X FASCIA

SECTION

SPACERS AS REQ'D @ FASTENERS

(A) FASCIA FLUSH W/DECKING

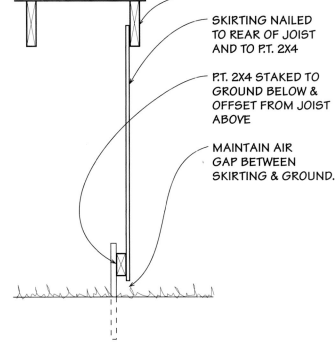

JOIST

SKIRTING NAILED TO REAR OF JOIST AND TO P.T. 2X4

P.T. 2X4 STAKED TO GROUND BELOW & OFFSET FROM JOIST ABOVE

MAINTAIN AIR GAP BETWEEN SKIRTING & GROUND.

(B) SKIRT ATTACHED TO JOISTS
PARALLEL TO JOISTS

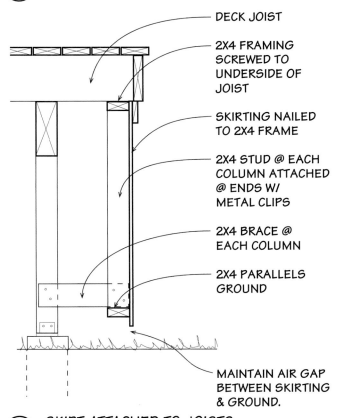

DECK JOIST

2X4 FRAMING SCREWED TO UNDERSIDE OF JOIST

SKIRTING NAILED TO 2X4 FRAME

2X4 STUD @ EACH COLUMN ATTACHED @ ENDS W/ METAL CLIPS

2X4 BRACE @ EACH COLUMN

2X4 PARALLELS GROUND

MAINTAIN AIR GAP BETWEEN SKIRTING & GROUND.

(C) SKIRT ATTACHED TO JOISTS
PERPENDICULAR TO JOISTS

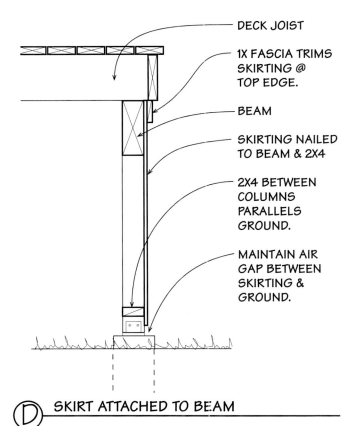

DECK JOIST

1X FASCIA TRIMS SKIRTING @ TOP EDGE.

BEAM

SKIRTING NAILED TO BEAM & 2X4

2X4 BETWEEN COLUMNS PARALLELS GROUND.

MAINTAIN AIR GAP BETWEEN SKIRTING & GROUND.

(D) SKIRT ATTACHED TO BEAM

NOTE: ADDITIONAL RAIL POSTS CAN BE BOLTED TO JOISTS FOR EXTRA SUPPORT.

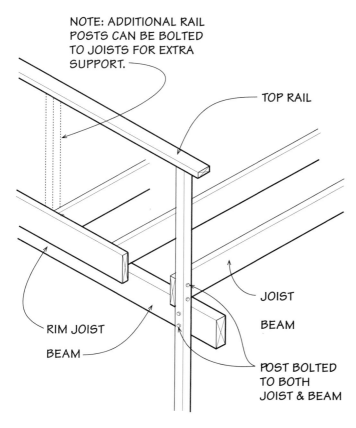

TOP RAIL

RIM JOIST

BEAM

JOIST

BEAM

POST BOLTED TO BOTH JOIST & BEAM

(A) CONTINUOUS POST GUARDRAIL

TOP RAIL

RAILING POST BOLTED TO RIM JOIST

METAL HANGER OR CLIP SECURES RIM JOIST TO DECK FRAMING.

BLOCKING TO STIFFEN RIM JOIST

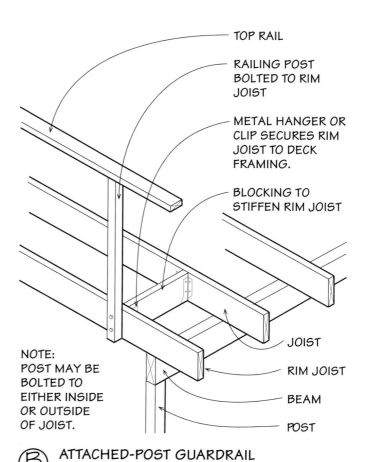

NOTE: POST MAY BE BOLTED TO EITHER INSIDE OR OUTSIDE OF JOIST.

JOIST

RIM JOIST

BEAM

POST

(B) ATTACHED-POST GUARDRAIL

GUARDRAILS—Guardrails are required at the edges of decks when the surface of the deck is 30 in. or more above the ground. Guardrails are like short fences and are supported by posts affixed to the deck at regular intervals. The rail design ranges from solid (like a wall) to transparent (like a window) to alternately solid and open (like a picket fence).

In every case, the building code governs the design of rails to prevent people from falling through or breaking them. The rail design must be tight enough to prevent a 4-in. sphere from passing through, and it must be able to withstand a substantial lateral load imposed from the inside (deck side) of the rail. Local codes vary on the specifics of these issues.

Posts—The stiffness of the posts that support the rail is the greatest challenge of guardrail design. One approach is to extend above deck level the same posts

that support the deck (see 116A). This strategy requires that the location of columns that support the deck coincide with the need for railing supports. The quality of material used for exposed railing supports also must be carried all the way to the foundation.

A second approach is to attach railing posts to deck framing—usually the joists (see 116B). This approach is more practical because it allows greater flexibility in the location of railing supports, but it is not as strong, and extra bracing of the joists is often required, especially if the railing is parallel to the joists.

Railings and features—Having established a strategy for the attachment of railing posts, there are a variety of ways in which the railing itself may be constructed (see 117A to D). Benches are also practical additions to the edges of decks, and in many jurisdictions, the code allows benches of certain types to be substituted where a railing would be required (see 118A, B, & C).

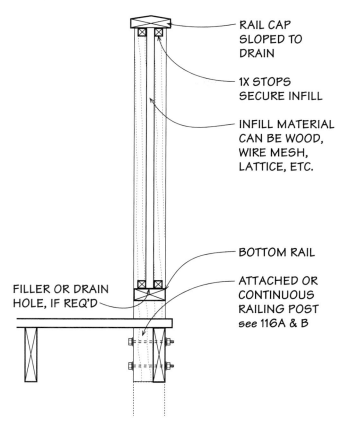

RAIL CAP SLOPED TO DRAIN

1X STOPS SECURE INFILL

INFILL MATERIAL CAN BE WOOD, WIRE MESH, LATTICE, ETC.

BOTTOM RAIL

ATTACHED OR CONTINUOUS RAILING POST
see 116A & B

FILLER OR DRAIN HOLE, IF REQ'D

Ⓐ OPEN RAILING W/ INFILL ASSEMBLY

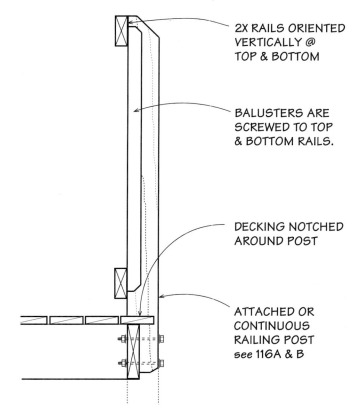

2X RAILS ORIENTED VERTICALLY @ TOP & BOTTOM

BALUSTERS ARE SCREWED TO TOP & BOTTOM RAILS.

DECKING NOTCHED AROUND POST

ATTACHED OR CONTINUOUS RAILING POST
see 116A & B

Ⓑ OPEN RAILING W/ OUTSIDE BALLUSTER

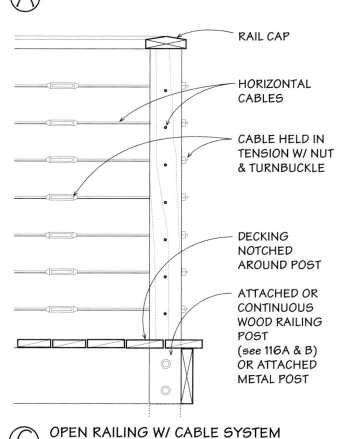

RAIL CAP

HORIZONTAL CABLES

CABLE HELD IN TENSION W/ NUT & TURNBUCKLE

DECKING NOTCHED AROUND POST

ATTACHED OR CONTINUOUS WOOD RAILING POST
(see 116A & B) OR ATTACHED METAL POST

Ⓒ OPEN RAILING W/ CABLE SYSTEM

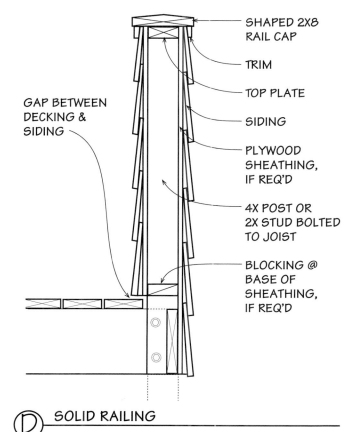

SHAPED 2X8 RAIL CAP

TRIM

TOP PLATE

SIDING

PLYWOOD SHEATHING, IF REQ'D

4X POST OR 2X STUD BOLTED TO JOIST

BLOCKING @ BASE OF SHEATHING, IF REQ'D

GAP BETWEEN DECKING & SIDING

Ⓓ SOLID RAILING

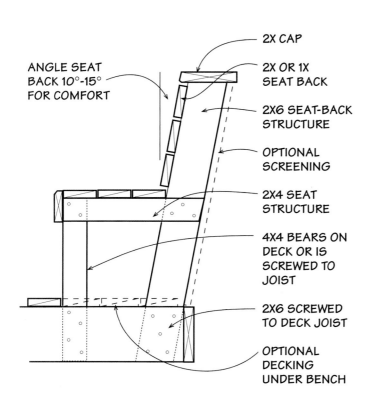

ANGLE SEAT BACK 10°-15° FOR COMFORT

2X CAP

2X OR 1X SEAT BACK

2X6 SEAT-BACK STRUCTURE

OPTIONAL SCREENING

2X4 SEAT STRUCTURE

4X4 BEARS ON DECK OR IS SCREWED TO JOIST

2X6 SCREWED TO DECK JOIST

OPTIONAL DECKING UNDER BENCH

(A) BUILT-IN SEAT W/BACK
SUPPORT BY JOISTS

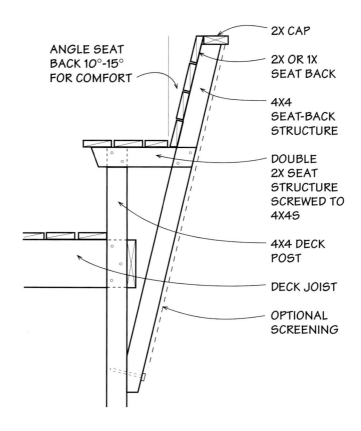

ANGLE SEAT BACK 10°-15° FOR COMFORT

2X CAP

2X OR 1X SEAT BACK

4X4 SEAT-BACK STRUCTURE

DOUBLE 2X SEAT STRUCTURE SCREWED TO 4X4S

4X4 DECK POST

DECK JOIST

OPTIONAL SCREENING

(B) BUILT-IN SEAT W/BACK
SUPPORT BY POSTS

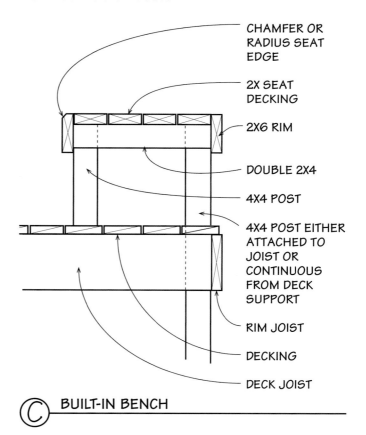

CHAMFER OR RADIUS SEAT EDGE

2X SEAT DECKING

2X6 RIM

DOUBLE 2X4

4X4 POST

4X4 POST EITHER ATTACHED TO JOIST OR CONTINUOUS FROM DECK SUPPORT

RIM JOIST

DECKING

DECK JOIST

(C) BUILT-IN BENCH

DECK STAIRS—Most decks require a few stairs to reach the level of the ground. Stairs also may descend to a deck from a porch or other level higher than the deck. Multilevel decks require stairs between levels.

Code requirements—All stairways are governed by building codes. The primary design issues addressed by codes are the configuration of the stair (rise, run, and width) and handrail design.

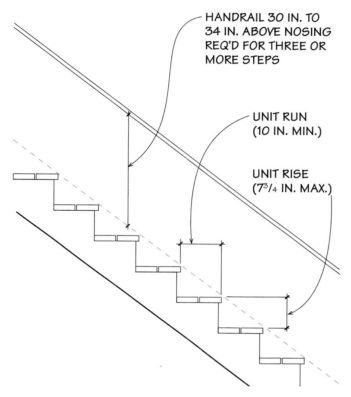

HANDRAIL 30 IN. TO 34 IN. ABOVE NOSING REQ'D FOR THREE OR MORE STEPS

UNIT RUN (10 IN. MIN.)

UNIT RISE (7³/₄ IN. MAX.)

Codes allow stairways to be steeper and narrower than is typically desired in an outdoor location. The rise and run of a stair is specified by code to be: 7¾ in. (max.) rise and 10 in. (min.) run. Outdoor steps are usually more relaxed than these proportions, however, because the human stride tends to stretch outdoors. Common proportions are a 6-in. rise with a 14-in. run. The minimum width of stairways is 36 in., but 48 in. and wider is much more common outdoors.

Handrails are required by code when there are three or more risers. When finished grade is 30 in. or more below the stairway, the handrail assembly must include balusters or other screening devices to prevent people from falling off the stair.

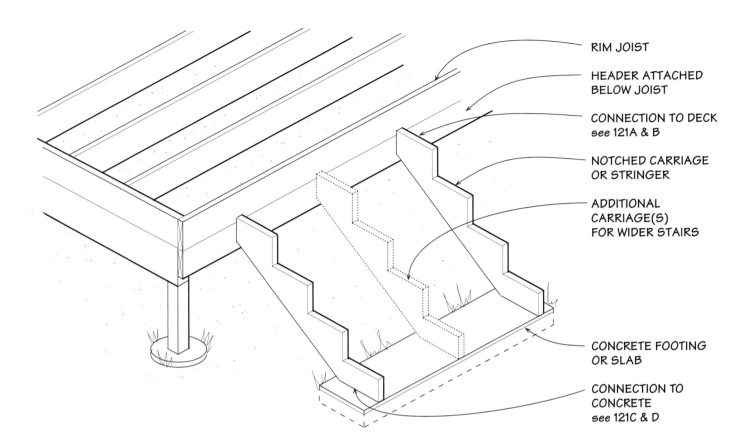

RIM JOIST

HEADER ATTACHED
BELOW JOIST

CONNECTION TO DECK
see 121A & B

NOTCHED CARRIAGE
OR STRINGER

ADDITIONAL
CARRIAGE(S)
FOR WIDER STAIRS

CONCRETE FOOTING
OR SLAB

CONNECTION TO
CONCRETE
see 121C & D

Carriage—The primary structural component of a stairway that spans from the top to the bottom of the stair is called a carriage (or stringer, jack, or buck, depending on where you live and who you're talking to).

The most common carriage is made from a 2x12 that is notched to the profile of the stairs. The notching considerably weakens the structural capacity of the carriage and limits its use to short spans (of 4 ft. to 5 ft.). Longer spans can be made with unnotched carriages. However, this type of carriage can be used only at the side of the stairway, so the width of the stairway is limited by the strength of the treads.

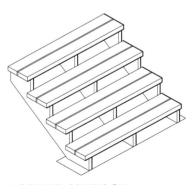

NOTCHED CARRIAGE

UNNOTCHED CARRIAGE

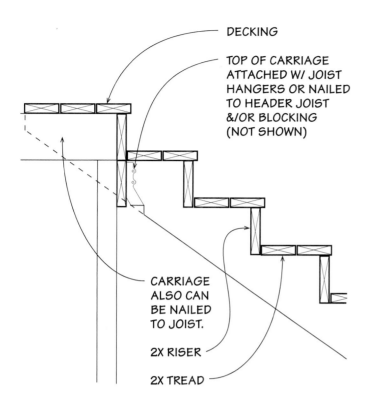

DECKING

TOP OF CARRIAGE ATTACHED W/ JOIST HANGERS OR NAILED TO HEADER JOIST &/OR BLOCKING (NOT SHOWN)

CARRIAGE ALSO CAN BE NAILED TO JOIST.

2X RISER

2X TREAD

(A) CARRIAGE @ DECK
NOTCHED CARRIAGE

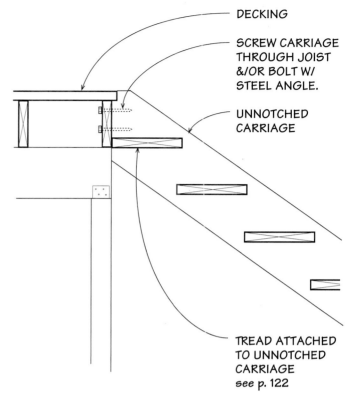

DECKING

SCREW CARRIAGE THROUGH JOIST &/OR BOLT W/ STEEL ANGLE.

UNNOTCHED CARRIAGE

TREAD ATTACHED TO UNNOTCHED CARRIAGE
see p. 122

(B) CARRIAGE @ DECK
UNNOTCHED CARRIAGE

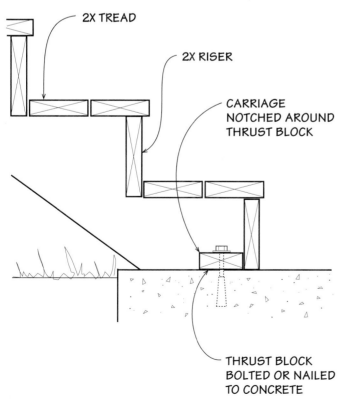

2X TREAD

2X RISER

CARRIAGE NOTCHED AROUND THRUST BLOCK

THRUST BLOCK BOLTED OR NAILED TO CONCRETE

(C) CARRIAGE CONNECTION @ BASE
NOTCHED CARRIAGE

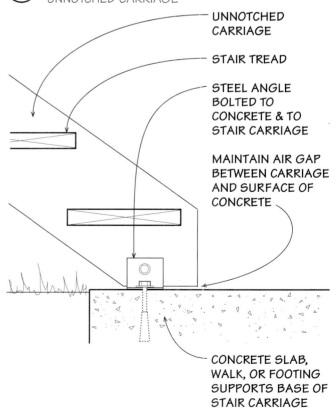

UNNOTCHED CARRIAGE

STAIR TREAD

STEEL ANGLE BOLTED TO CONCRETE & TO STAIR CARRIAGE

MAINTAIN AIR GAP BETWEEN CARRIAGE AND SURFACE OF CONCRETE

CONCRETE SLAB, WALK, OR FOOTING SUPPORTS BASE OF STAIR CARRIAGE

(D) CARRIAGE CONNECTION @ BASE

UNNOTCHED CARRIAGE

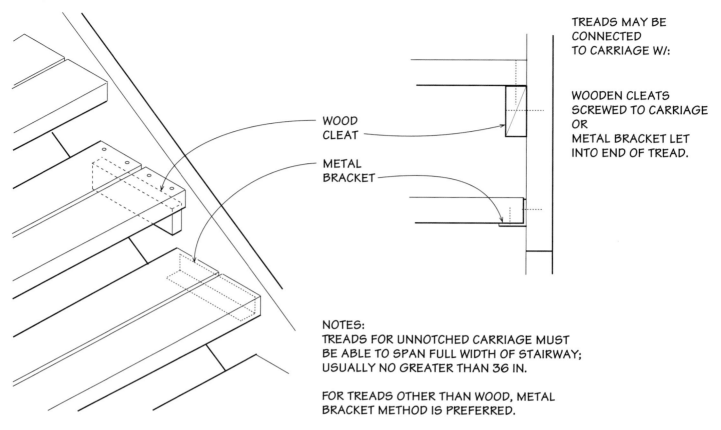

WOOD
CLEAT

METAL
BRACKET

TREADS MAY BE
CONNECTED
TO CARRIAGE W/:

WOODEN CLEATS
SCREWED TO CARRIAGE
OR
METAL BRACKET LET
INTO END OF TREAD.

NOTES:
TREADS FOR UNNOTCHED CARRIAGE MUST
BE ABLE TO SPAN FULL WIDTH OF STAIRWAY;
USUALLY NO GREATER THAN 36 IN.

FOR TREADS OTHER THAN WOOD, METAL
BRACKET METHOD IS PREFERRED.

(A) CONNECTION OF TREADS TO UNNOTCHED CARRIAGE

Treads and risers—The treads and risers of most wooden outdoor stairs are made with 2x material—usually 2x4 or 2x6 to minimize cupping. The ability of 2x4 and 2x6 boards to span between carriages is limited to about 24 in., so carriages are typically spaced at this distance or closer and are concealed by the treads and risers.

Occasionally, it is desirable to eliminate the risers in order to see through the stair. In this case, stiff treads that can span between unnotched carriages are an option (see 122A). Treads made of metal or precast concrete are ideal for this purpose because they are strong and will not cup or otherwise distort with moisture.

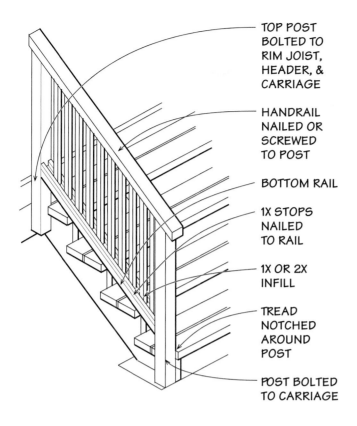

TOP POST
BOLTED TO
RIM JOIST,
HEADER, &
CARRIAGE

HANDRAIL
NAILED OR
SCREWED
TO POST

BOTTOM RAIL

1X STOPS
NAILED
TO RAIL

1X OR 2X
INFILL

TREAD
NOTCHED
AROUND
POST

POST BOLTED
TO CARRIAGE

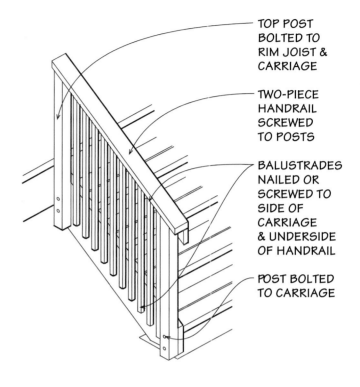

TOP POST
BOLTED TO
RIM JOIST &
CARRIAGE

TWO-PIECE
HANDRAIL
SCREWED
TO POSTS

BALUSTRADES
NAILED OR
SCREWED TO
SIDE OF
CARRIAGE
& UNDERSIDE
OF HANDRAIL

POST BOLTED
TO CARRIAGE

 HANDRAIL FOR NOTCHED CARRIAGE

 HANDRAIL FOR UNNOTCHED CARRIAGE

Handrails—The code prescribes handrails for most exterior stairs that rise 30 in. or more above the ground. In addition to providing a grip, the handrail prevents children from falling off the side of the stair. Handrails are typically supported either by newel posts attached to the deck framing or by balusters attached to the carriage (see 112A & B). Occasionally, a handrail is made as a solid wall with siding to match the house (see 117D).

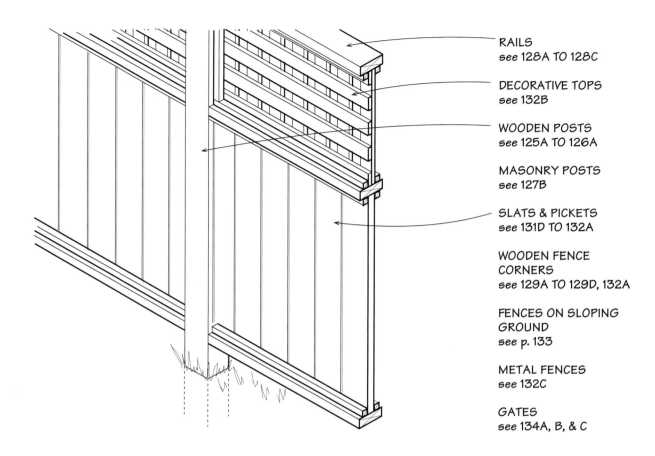

RAILS
see 128A TO 128C

DECORATIVE TOPS
see 132B

WOODEN POSTS
see 125A TO 126A

MASONRY POSTS
see 127B

SLATS & PICKETS
see 131D TO 132A

WOODEN FENCE
CORNERS
see 129A TO 129D, 132A

FENCES ON SLOPING
GROUND
see p. 133

METAL FENCES
see 132C

GATES
see 134A, B, & C

FENCES

Fences are prevalent in residential landscapes, where they provide security, privacy, and control of children and animals. Ranging from the simple and humble wooden picket fence to elaborate composite structures made of masonry, wood, and steel, fences are built by contractors and homeowners alike. Despite their range of purpose and style, all fences share a number of basic design issues.

Structurally, fences are distinct from other landscape structures in that lateral loads of wind, leaning people and animals, and windblown snow are more important to consider than vertical (gravitational) loads. In fact, fences are unlikely to have any vertical loads greater than their own weight. Lateral loads can be considerable, so the posts that provide structural stability to fences are generally rooted deeply into the ground (see the diagram at right).

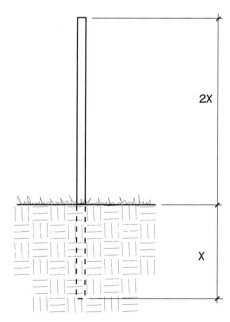

DEPTH OF POST EMBEDMENT DEPENDS ON A NUMBER OF FACTORS, INCLUDING SOIL TYPE. GENERALLY, 1 FT. OF EMBEDMENT FOR EVERY 2 FT. ABOVE GROUND IS SUFFICIENT.

The fence's appearance is usually important, and a number of issues must be considered in this regard. Is the view of the fence from one side more important than from the other? If so, what are the aesthetic criteria from the less important side? How does the top of the fence meet the sky? How will weathering affect the appearance of the fence?

When answers to the aesthetic questions are combined with the realities of technical factors such as the construction of gates, details at corners, and the adaptation of the fence to changes in grade, a comprehensive design can be made. Although the range of aesthetic choices is virtually limitless, the solidity and durability of a fence design ultimately depend on an understanding of technical factors.

FENCE POSTS—The fence post is the key to a stable and durable fence. Only when fence posts remain vertical and do not settle will the rest of the fence remain true and straight, allowing the joints between its pieces to stay unstressed.

Set and driven posts—Fence posts are typically set into the ground to achieve their stability. Holes are dug and the posts are set in them and braced. The holes are then backfilled with soil, gravel, or concrete in order to lock the post into its permanent position (see 125A & B). Sometimes a mixture of soil and gravel is the most effective backfill material. When soil and/or gravel are used, the material is tamped to compact it between the post and the sides of the hole. When concrete is used as backfill, the hole does not need to be quite as deep or wide because the concrete is stronger and denser and does not require compaction. The choice of backfill materials depends on a number of factors, including soil characteristics, fence type, and cost.

In cold regions, the heaving of frozen ground in winter can play havoc with fence posts. To minimize problems with freezing, post holes should be excavated to a depth below the frost line, and the post should be anchored to the soil below the frost line so that expansion of freezing soil will not lift the post.

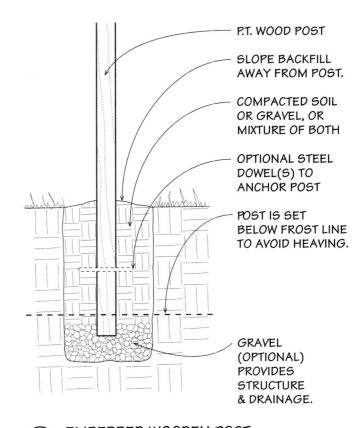

- P.T. WOOD POST
- SLOPE BACKFILL AWAY FROM POST.
- COMPACTED SOIL OR GRAVEL, OR MIXTURE OF BOTH
- OPTIONAL STEEL DOWEL(S) TO ANCHOR POST
- POST IS SET BELOW FROST LINE TO AVOID HEAVING.
- GRAVEL (OPTIONAL) PROVIDES STRUCTURE & DRAINAGE.

Ⓐ **EMBEDDED WOODEN POST**
SOIL &/OR GRAVEL BACKFILL

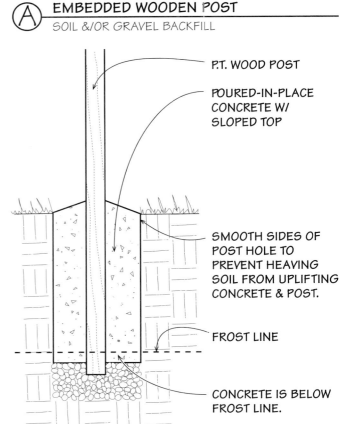

- P.T. WOOD POST
- POURED-IN-PLACE CONCRETE W/ SLOPED TOP
- SMOOTH SIDES OF POST HOLE TO PREVENT HEAVING SOIL FROM UPLIFTING CONCRETE & POST.
- FROST LINE
- CONCRETE IS BELOW FROST LINE.

Ⓑ **EMBEDDED WOODEN POST**
CONCRETE BACKFILL

With ideal soil conditions, fence posts can be driven into the ground. Typically, the wooden or metal post has a pointed end, and a weighted tube or post driver is placed over it. The post driver is repeatedly lifted and dropped to drive the post into the ground. The post compacts the soil around it as it is driven into the ground. Even in the best conditions, it is virtually impossible to achieve true verticality of driven posts, so it is best to reserve this labor-saving technique for fences that can be rather loose in their construction and appearance.

The principal drawback of setting or driving wooden posts into the ground is their tendency to rot because of constant exposure to moisture. The use of decay-resistant species or preservative-treated wood is the main defense against rot, but extra preservative and sealant is often applied to the zone just above and below the ground level where almost constant moisture in the presence of oxygen makes the posts most vulnerable.

The life of wooden posts can be prolonged by providing good drainage at the base of the post holes and by sealing the tops of the holes with concrete to minimize storm water infiltration. Using metal caps or sloping the tops of wooden posts will minimize rot at the top end of the post (see 126A).

Bolted posts—Where a fence is to be built on a concrete or masonry retaining wall, slab, or other stable structure, or where the ground is too hard for holes to be dug, fence posts can be bolted to metal brackets firmly embedded in concrete. The brackets must be tall enough and the bolts spaced apart far enough to resist lateral forces on the fence (see 127A).

Columns—Fences also may be stabilized with masonry or concrete columns, which do not have any of the decay problems associated with wood (and metal) (see 127B). Masonry and concrete columns are very common in regions such as the Southwest where there is virtually no frost to heave the soil, so the footings for the columns may be constructed near the surface of the ground. The critical factor for the stability of a masonry or concrete column is the width of the footing that prevents overturning.

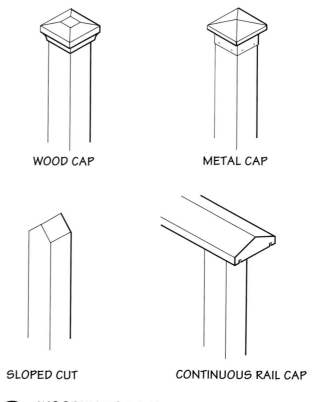

WOOD CAP   METAL CAP

SLOPED CUT   CONTINUOUS RAIL CAP

 WOODEN POST TOPS

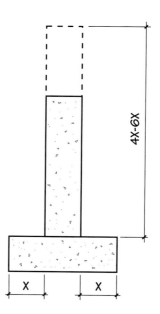

4X–6X

X   X

RAILS—Most fences have at least two horizontal rails that provide support for vertical slats, pickets, or lattice panels. Additional rails can be added to reinforce vertical slats, helping to prevent warping, or so that more than one type of slat or panel can be attached to the same fence. For example, lattice can be added to the top of a board fence.

Rails are usually made of 2x4 decay-resistant or preservative-treated wood. Orientation and attachment of the rails to the posts depend on concerns about structure, weathering, and the style of fence (see 128A to C). The relationship of the rails to the posts will impact the design at fence corners (see 129 A to D).

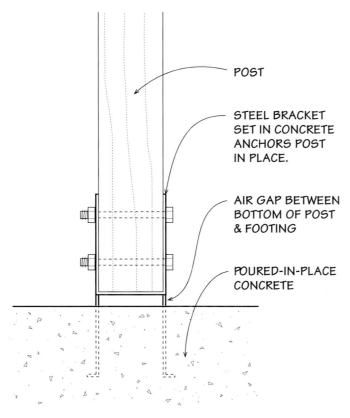

POST

STEEL BRACKET SET IN CONCRETE ANCHORS POST IN PLACE.

AIR GAP BETWEEN BOTTOM OF POST & FOOTING

POURED-IN-PLACE CONCRETE

(A) WOODEN POST W/ BRACKET

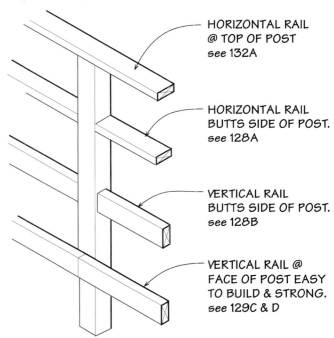

HORIZONTAL RAIL @ TOP OF POST
see 132A

HORIZONTAL RAIL BUTTS SIDE OF POST.
see 128A

VERTICAL RAIL BUTTS SIDE OF POST.
see 128B

VERTICAL RAIL @ FACE OF POST EASY TO BUILD & STRONG.
see 129C & D

NOTE:
VERTICAL RAIL ORIENTATION PROVIDES STRENGTH TO RESIST SAGGING. HORIZONTAL RAIL ORIENTATION PROVIDES STRENGTH AGAINST WIND & OTHER LATERAL LOADS.

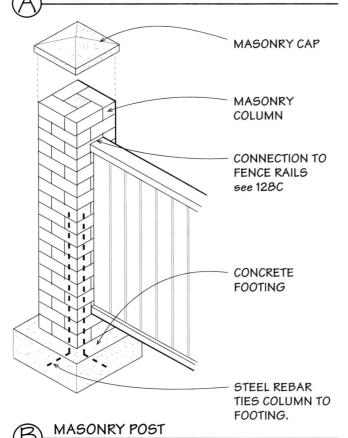

MASONRY CAP

MASONRY COLUMN

CONNECTION TO FENCE RAILS
see 128C

CONCRETE FOOTING

STEEL REBAR TIES COLUMN TO FOOTING.

(B) MASONRY POST

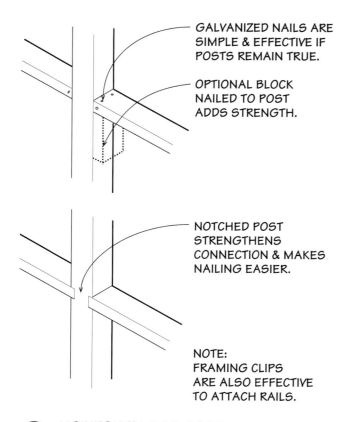

GALVANIZED NAILS ARE
SIMPLE & EFFECTIVE IF
POSTS REMAIN TRUE.

OPTIONAL BLOCK
NAILED TO POST
ADDS STRENGTH.

NOTCHED POST
STRENGTHENS
CONNECTION & MAKES
NAILING EASIER.

NOTE:
FRAMING CLIPS
ARE ALSO EFFECTIVE
TO ATTACH RAILS.

Ⓐ HORIZONTAL RAIL POST

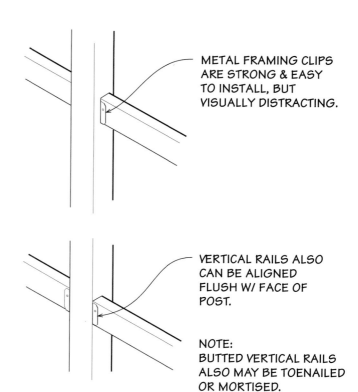

METAL FRAMING CLIPS
ARE STRONG & EASY
TO INSTALL, BUT
VISUALLY DISTRACTING.

VERTICAL RAILS ALSO
CAN BE ALIGNED
FLUSH W/ FACE OF
POST.

NOTE:
BUTTED VERTICAL RAILS
ALSO MAY BE TOENAILED
OR MORTISED.

Ⓑ VERTICAL RAIL POST
BUTTED RAIL

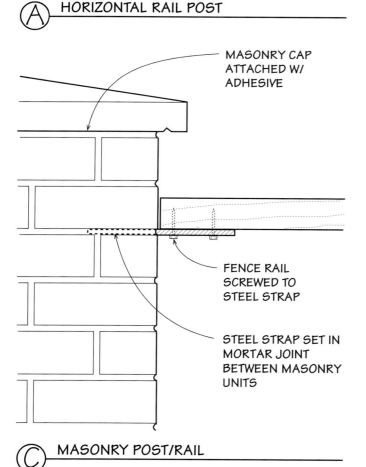

MASONRY CAP
ATTACHED W/
ADHESIVE

FENCE RAIL
SCREWED TO
STEEL STRAP

STEEL STRAP SET IN
MORTAR JOINT
BETWEEN MASONRY
UNITS

Ⓒ MASONRY POST/RAIL

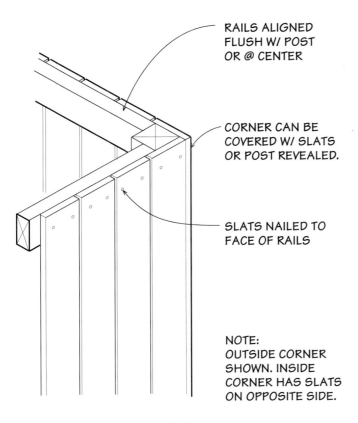

RAILS ALIGNED FLUSH W/ POST OR @ CENTER

CORNER CAN BE COVERED W/ SLATS OR POST REVEALED.

SLATS NAILED TO FACE OF RAILS

NOTE: OUTSIDE CORNER SHOWN. INSIDE CORNER HAS SLATS ON OPPOSITE SIDE.

Ⓐ **VERTICAL BUTTED RAILS**
INSIDE OR OUTSIDE CORNER

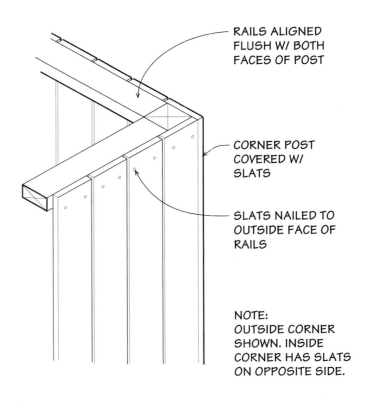

RAILS ALIGNED FLUSH W/ BOTH FACES OF POST

CORNER POST COVERED W/ SLATS

SLATS NAILED TO OUTSIDE FACE OF RAILS

NOTE: OUTSIDE CORNER SHOWN. INSIDE CORNER HAS SLATS ON OPPOSITE SIDE.

Ⓑ **HORIZONTAL BUTTED RAILS**
INSIDE OR OUTSIDE CORNER

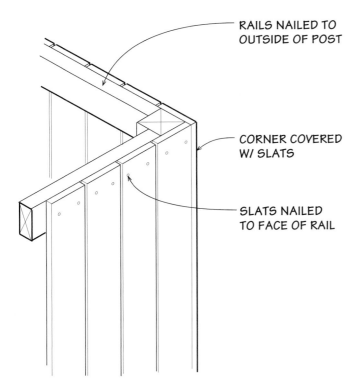

RAILS NAILED TO OUTSIDE OF POST

CORNER COVERED W/ SLATS

SLATS NAILED TO FACE OF RAIL

Ⓒ **FACE-MOUNTED RAILS**
OUTSIDE CORNER

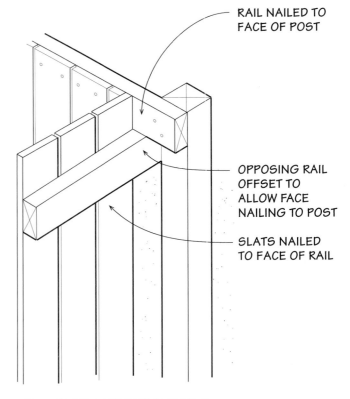

RAIL NAILED TO FACE OF POST

OPPOSING RAIL OFFSET TO ALLOW FACE NAILING TO POST

SLATS NAILED TO FACE OF RAIL

Ⓓ **FACE-MOUNTED RAILS**
INSIDE CORNER

SLATS, PICKETS, LATTICE, AND OTHER PANELS—Because fences are such a visible component of the landscape, their designers usually try to make them visually appealing. This is accomplished by introducing texture and pattern into the surface of the fence by manipulating the size, spacing, and orientation of the slats, pickets, lattice, and other materials that are attached to the framework of posts and rails to complete the fence.

These materials, generally made of wood, are usually oriented vertically and attached to the rails (see 130A to 131A), but they can be oriented in other ways and can be attached in a variety of patterns (see 131B). In some cases, horizontal boards can be used to eliminate the need for rails altogether (see 131C). Prefabricated panels (4 ft., 5 ft., or 6 ft. high, and 8 ft. long) can also be attached directly to posts, eliminating the need for rails or site-applied slats. In addition, the top of the fence is often given special treatment to make it tidy-looking (see 132A) or decorative (see 132B).

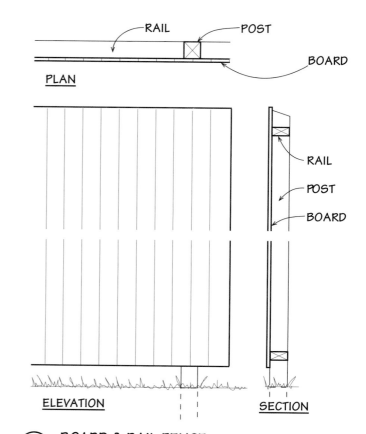

Ⓐ **BOARD & RAIL FENCE**

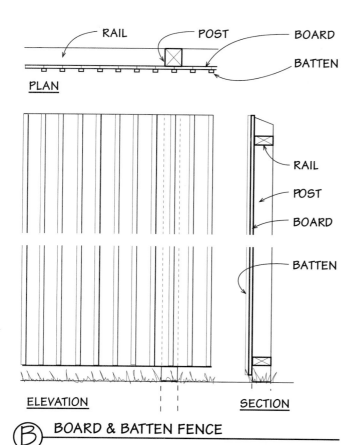

Ⓑ **BOARD & BATTEN FENCE**

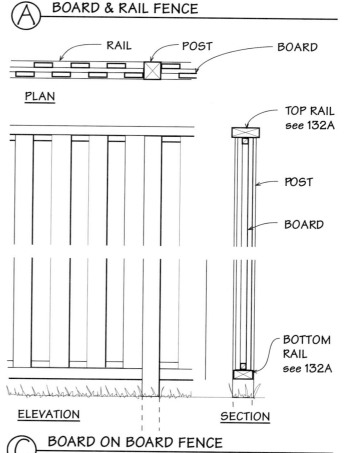

Ⓒ **BOARD ON BOARD FENCE**

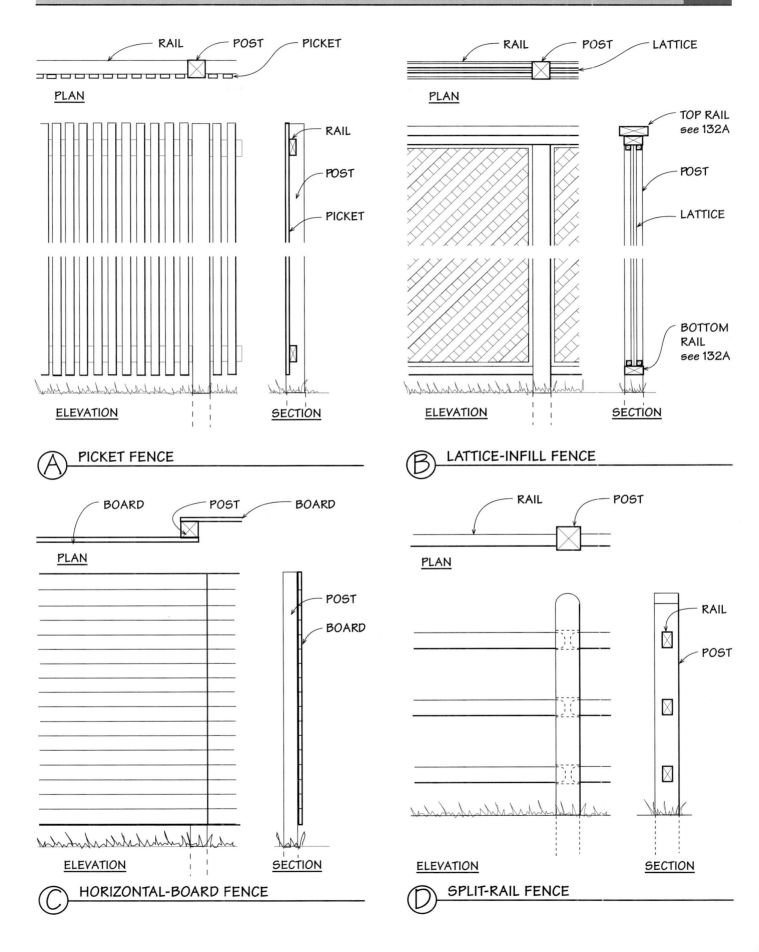

RAIL — POST — PICKET

PLAN

RAIL

POST

PICKET

ELEVATION   SECTION

Ⓐ PICKET FENCE

RAIL — POST — LATTICE

PLAN

TOP RAIL
see 132A

POST

LATTICE

BOTTOM
RAIL
see 132A

ELEVATION   SECTION

Ⓑ LATTICE-INFILL FENCE

BOARD — POST — BOARD

PLAN

POST

BOARD

ELEVATION   SECTION

Ⓒ HORIZONTAL-BOARD FENCE

RAIL — POST

PLAN

RAIL

POST

ELEVATION   SECTION

Ⓓ SPLIT-RAIL FENCE

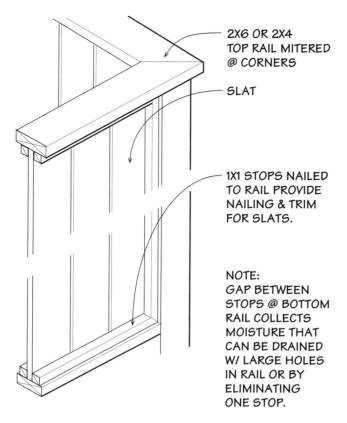

2X6 OR 2X4 TOP RAIL MITERED @ CORNERS

SLAT

1X1 STOPS NAILED TO RAIL PROVIDE NAILING & TRIM FOR SLATS.

NOTE: GAP BETWEEN STOPS @ BOTTOM RAIL COLLECTS MOISTURE THAT CAN BE DRAINED W/ LARGE HOLES IN RAIL OR BY ELIMINATING ONE STOP.

Ⓐ FINISHED TOP & BOTTOM RAILS

TOP RAIL OF DECORATIVE PORTION OF FENCE

DECORATIVE LATTICE OR SLAT PANEL ATTACHED TO RAILS W/ 1X STOPS

TOP RAIL OF MAIN FENCE

NOTE: RAILS W/ DADO ALLOW ELIMINATION OF STOPS.

Ⓑ DECORATIVE FENCE TOP

CHAIN-LINK FENCES—Chain-link fences, most commonly found in institutional settings, are also used to define residential yards. These fences are long-lasting, secure, and easily adapted to sloping terrain. In addition, they are relatively transparent compared to other fences.

The frame for a chain-link fence is made of galvanized tubing (pipe) that is connected with specialized fittings. The metal posts are set into concrete because they are slender and need extra reinforcement to be rigid. The chain link mesh is stretched over the metal framework and attached with wire clips and other devices (see 132C).

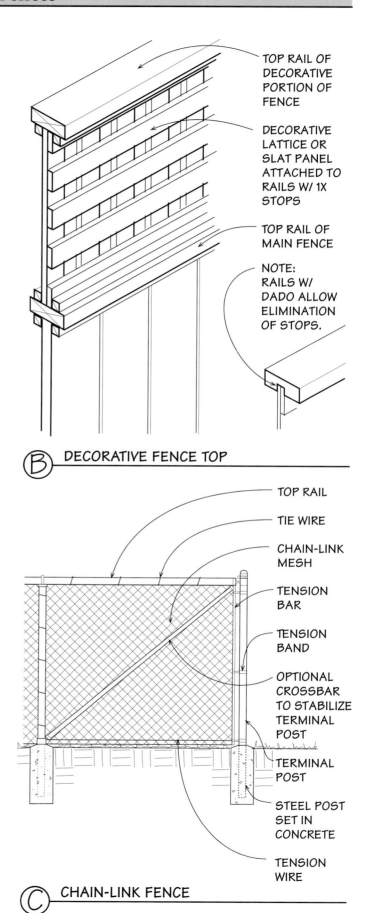

TOP RAIL

TIE WIRE

CHAIN-LINK MESH

TENSION BAR

TENSION BAND

OPTIONAL CROSSBAR TO STABILIZE TERMINAL POST

TERMINAL POST

STEEL POST SET IN CONCRETE

TENSION WIRE

Ⓒ CHAIN-LINK FENCE

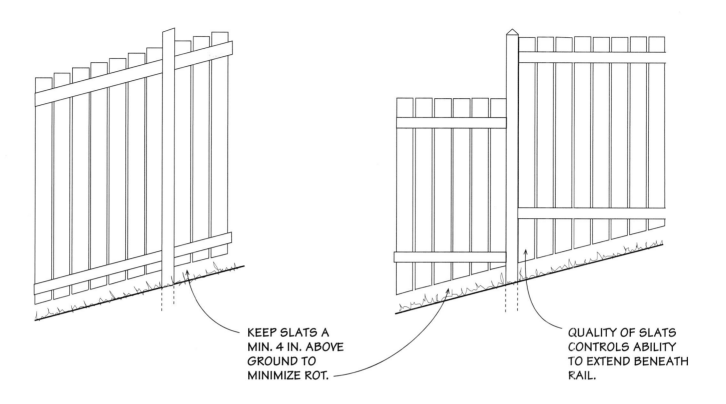

KEEP SLATS A
MIN. 4 IN. ABOVE
GROUND TO
MINIMIZE ROT.

QUALITY OF SLATS
CONTROLS ABILITY
TO EXTEND BENEATH
RAIL.

CONTOURING—It is very rare that a fence is built on dead-level ground. Therefore, a strategy must be employed to match the regular structure of the fence to the irregularities of the ground. There are two basic techniques for this purpose—make the top of the fence parallel to the ground or keep the top of the fence level, stepping it when required to maintain a minimum fence height (see above). The former is usually less expensive and requires fewer materials than does the latter. Simple fence styles with vertical boards generally work best for both strategies.

GATES—Gates are designed and constructed in a similar fashion to fences but have unique requirements to withstand repetitive opening and closing. The two fence posts that are adjacent to a gate must be stronger than typical posts because the gate hangs from one post and slams against the other. To accomplish this, posts at gates are often oversized and set deep into the ground.

Gates are similar in their structural requirements to doors. The frame of the gate must be sturdy at the corners to withstand the twisting (torque) action of opening and closing, and it must be braced diagonally to resist sagging (see 134B & C).

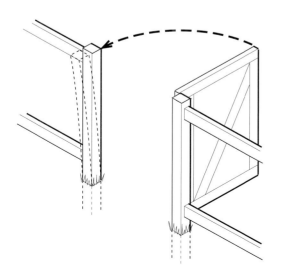

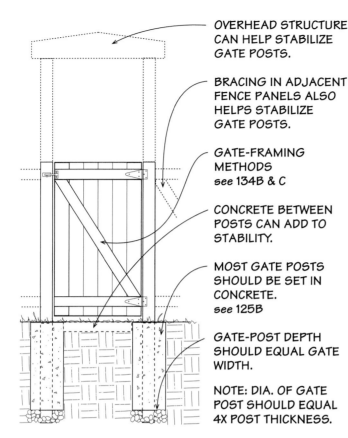

OVERHEAD STRUCTURE CAN HELP STABILIZE GATE POSTS.

BRACING IN ADJACENT FENCE PANELS ALSO HELPS STABILIZE GATE POSTS.

GATE-FRAMING METHODS
see 134B & C

CONCRETE BETWEEN POSTS CAN ADD TO STABILITY.

MOST GATE POSTS SHOULD BE SET IN CONCRETE.
see 125B

GATE-POST DEPTH SHOULD EQUAL GATE WIDTH.

NOTE: DIA. OF GATE POST SHOULD EQUAL 4X POST THICKNESS.

(A) GATES

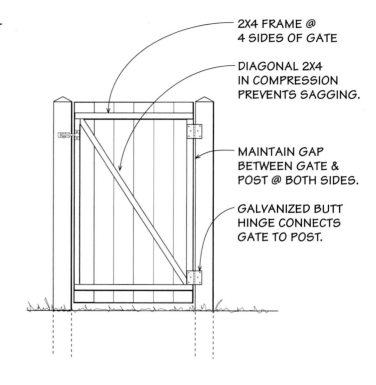

2X4 FRAME @ 4 SIDES OF GATE

DIAGONAL 2X4 IN COMPRESSION PREVENTS SAGGING.

MAINTAIN GAP BETWEEN GATE & POST @ BOTH SIDES.

GALVANIZED BUTT HINGE CONNECTS GATE TO POST.

(B) FRAME GATE

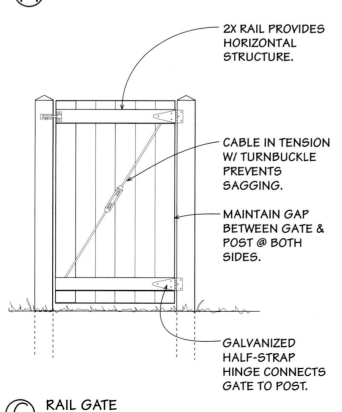

2X RAIL PROVIDES HORIZONTAL STRUCTURE.

CABLE IN TENSION W/ TURNBUCKLE PREVENTS SAGGING.

MAINTAIN GAP BETWEEN GATE & POST @ BOTH SIDES.

GALVANIZED HALF-STRAP HINGE CONNECTS GATE TO POST.

(C) RAIL GATE

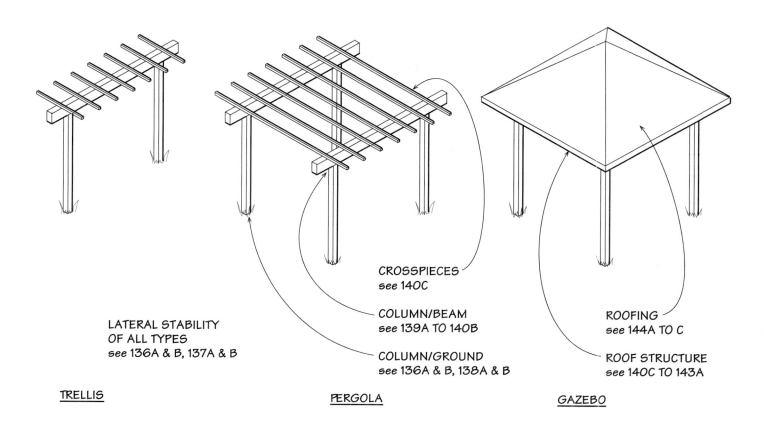

LATERAL STABILITY
OF ALL TYPES
see 136A & B, 137A & B

CROSSPIECES
see 140C

COLUMN/BEAM
see 139A TO 140B

COLUMN/GROUND
see 136A & B, 138A & B

ROOFING
see 144A TO C

ROOF STRUCTURE
see 140C TO 143A

TRELLIS

PERGOLA

GAZEBO

OUTBUILDINGS AND GARDEN STRUCTURES

There are many types of structures that can be built to add utility and beauty to a garden. Trellises, arbors, and pergolas are all open structures used to support vines and other climbing plants. Gazebos and belvederes are roofed structures that provide shelter in the garden. All can be elegant visual counterpoints to the planted landscape while serving practical functions as well.

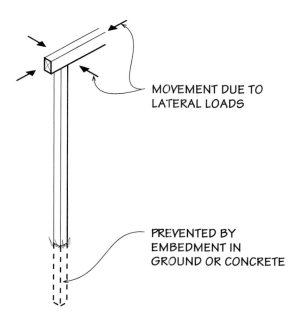

MOVEMENT DUE TO
LATERAL LOADS

PREVENTED BY
EMBEDMENT IN
GROUND OR CONCRETE

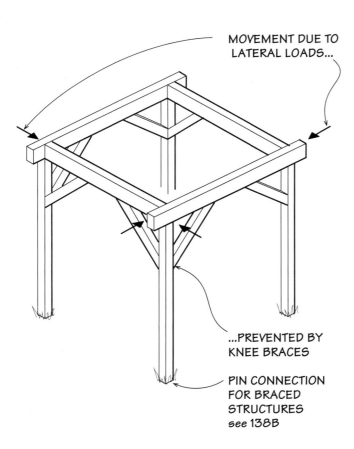

MOVEMENT DUE TO LATERAL LOADS...

...PREVENTED BY KNEE BRACES

PIN CONNECTION FOR BRACED STRUCTURES
see 138B

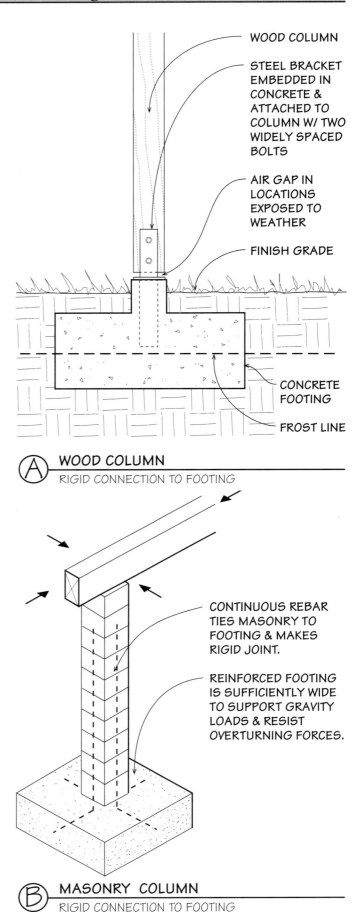

WOOD COLUMN

STEEL BRACKET EMBEDDED IN CONCRETE & ATTACHED TO COLUMN W/ TWO WIDELY SPACED BOLTS

AIR GAP IN LOCATIONS EXPOSED TO WEATHER

FINISH GRADE

CONCRETE FOOTING

FROST LINE

(A) WOOD COLUMN
RIGID CONNECTION TO FOOTING

CONTINUOUS REBAR TIES MASONRY TO FOOTING & MAKES RIGID JOINT.

REINFORCED FOOTING IS SUFFICIENTLY WIDE TO SUPPORT GRAVITY LOADS & RESIST OVERTURNING FORCES.

(B) MASONRY COLUMN
RIGID CONNECTION TO FOOTING

STRUCTURAL SUPPORT—The structural support of these outbuildings is usually reasonably straightforward, but design decisions concerning the resistance of lateral forces are strongly tied to decisions about appearance. Small, lightweight structures such as trellises and arbors may be constructed on posts that are embedded into the ground like fence posts. The posts for larger and more complex structures are often stabilized by rigid connections to a spread footing at their base (see 136A & B, 138A). If diagonal braces are visually acceptable, they can be used between post and beam for lateral stability of most structures (see 136A & B).

WEATHER DETAILING—Most garden structures have members and joints that are exposed to the weather, and these should be designed to minimize deterioration due to the weather. The principles outlined in the introduction to this section should be reviewed for these cases (see 126A).

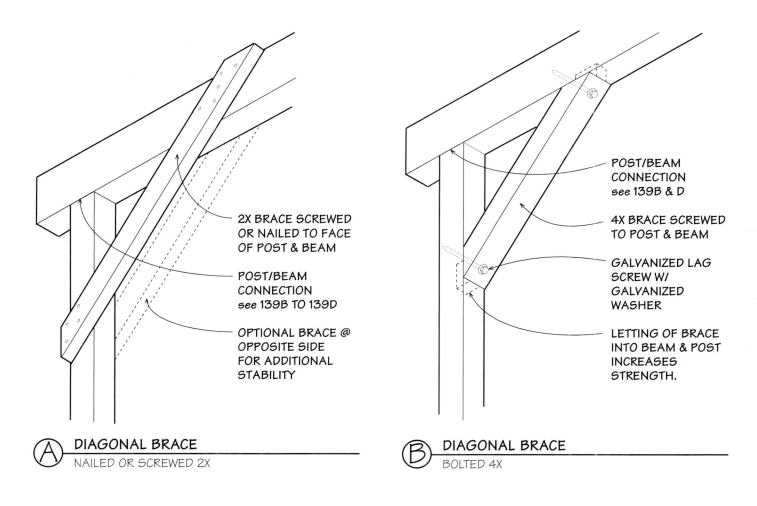

2X BRACE SCREWED
OR NAILED TO FACE
OF POST & BEAM

POST/BEAM
CONNECTION
see 139B TO 139D

OPTIONAL BRACE @
OPPOSITE SIDE
FOR ADDITIONAL
STABILITY

POST/BEAM
CONNECTION
see 139B & D

4X BRACE SCREWED
TO POST & BEAM

GALVANIZED LAG
SCREW W/
GALVANIZED
WASHER

LETTING OF BRACE
INTO BEAM & POST
INCREASES
STRENGTH.

(A) **DIAGONAL BRACE**
NAILED OR SCREWED 2X

(B) **DIAGONAL BRACE**
BOLTED 4X

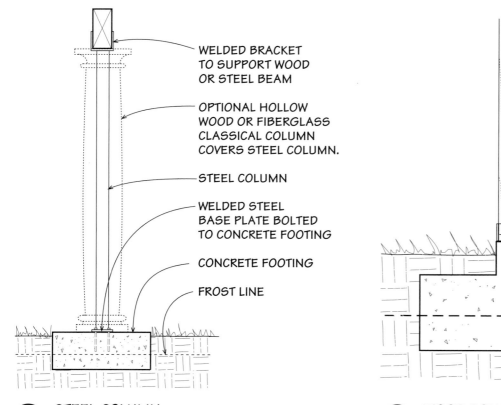

WELDED BRACKET
TO SUPPORT WOOD
OR STEEL BEAM

OPTIONAL HOLLOW
WOOD OR FIBERGLASS
CLASSICAL COLUMN
COVERS STEEL COLUMN.

STEEL COLUMN

WELDED STEEL
BASE PLATE BOLTED
TO CONCRETE FOOTING

CONCRETE FOOTING

FROST LINE

(A) **STEEL COLUMN**
RIGID CONNECTION TO FOOTING

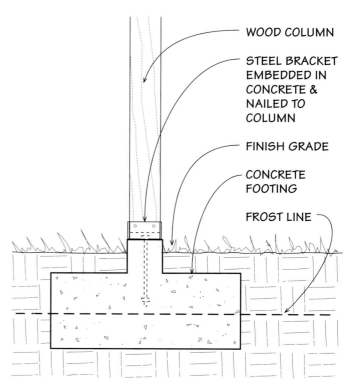

WOOD COLUMN

STEEL BRACKET
EMBEDDED IN
CONCRETE &
NAILED TO
COLUMN

FINISH GRADE

CONCRETE
FOOTING

FROST LINE

(B) **WOOD COLUMN**
PIN CONNECTION TO FOOTING

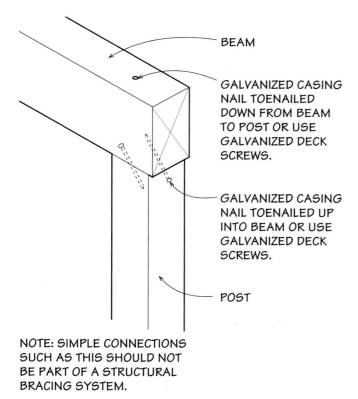

BEAM

GALVANIZED CASING NAIL TOENAILED DOWN FROM BEAM TO POST OR USE GALVANIZED DECK SCREWS.

GALVANIZED CASING NAIL TOENAILED UP INTO BEAM OR USE GALVANIZED DECK SCREWS.

POST

NOTE: SIMPLE CONNECTIONS SUCH AS THIS SHOULD NOT BE PART OF A STRUCTURAL BRACING SYSTEM.

(A) **POST/BEAM CONNECTION**
NAILED OR SCREWED

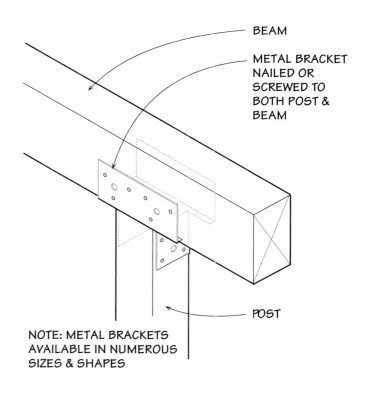

BEAM

METAL BRACKET NAILED OR SCREWED TO BOTH POST & BEAM

POST

NOTE: METAL BRACKETS AVAILABLE IN NUMEROUS SIZES & SHAPES

(B) **POST/BEAM CONNECTION**
METAL BRACKET

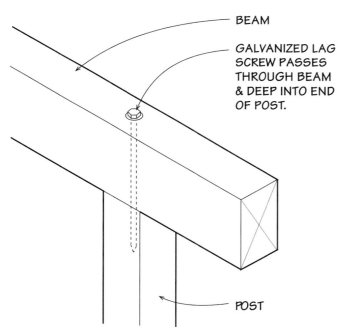

BEAM

GALVANIZED LAG SCREW PASSES THROUGH BEAM & DEEP INTO END OF POST.

POST

NOTE: LAG MAY BE COUNTERSUNK INTO BEAM IF PROTECTED FROM THE WEATHER.

(C) **POST/BEAM CONNECTION**
BOLTED FROM TOP

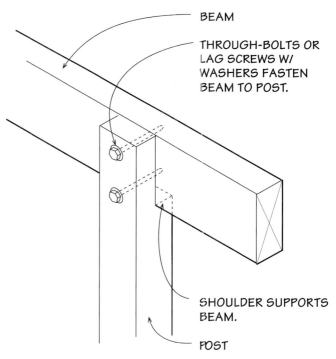

BEAM

THROUGH-BOLTS OR LAG SCREWS W/ WASHERS FASTEN BEAM TO POST.

SHOULDER SUPPORTS BEAM.

POST

A STRONG & ELEGANT CONNECTION, BEST USED WHEN PROTECTED FROM THE WEATHER.

(D) **POST/BEAM CONNECTION**
NOTCHED POST

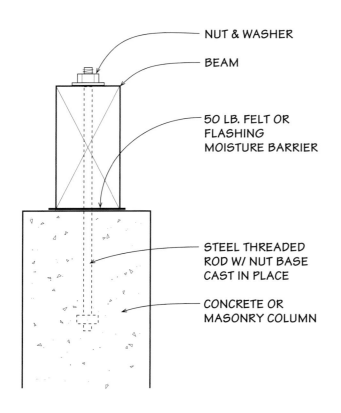

NUT & WASHER

BEAM

50 LB. FELT OR FLASHING MOISTURE BARRIER

STEEL THREADED ROD W/ NUT BASE CAST IN PLACE

CONCRETE OR MASONRY COLUMN

Ⓐ **CONCRETE OR MASONRY COLUMN/BEAM**
EMBEDDED BOLT

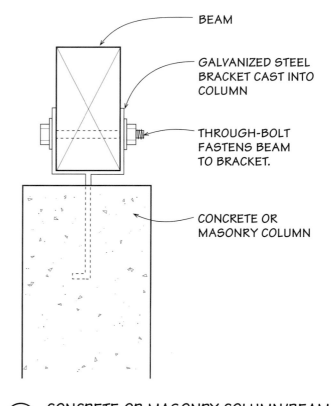

BEAM

GALVANIZED STEEL BRACKET CAST INTO COLUMN

THROUGH-BOLT FASTENS BEAM TO BRACKET.

CONCRETE OR MASONRY COLUMN

Ⓑ **CONCRETE OR MASONRY COLUMN/BEAM**
METAL BRACKET

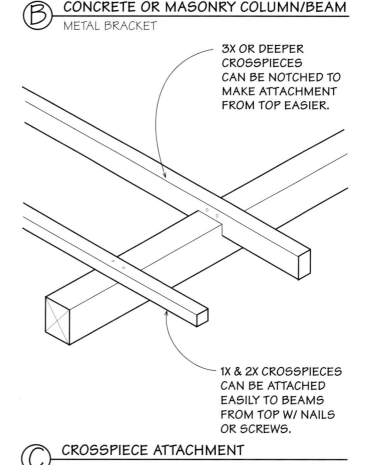

3X OR DEEPER CROSSPIECES CAN BE NOTCHED TO MAKE ATTACHMENT FROM TOP EASIER.

1X & 2X CROSSPIECES CAN BE ATTACHED EASILY TO BEAMS FROM TOP W/ NAILS OR SCREWS.

Ⓒ **CROSSPIECE ATTACHMENT**

ROOF STRUCTURE—Roofs for gazebos and belvederes provide both shade and weather protection for the area that they cover. Conveniently, the weathertight roof protects the structure that supports it.

There are three basic roof shapes, and it is the shape of the roof that determines its structure.

A simple shed roof is framed with parallel beams—one higher than the other—and rafters that span between them (see 142A).

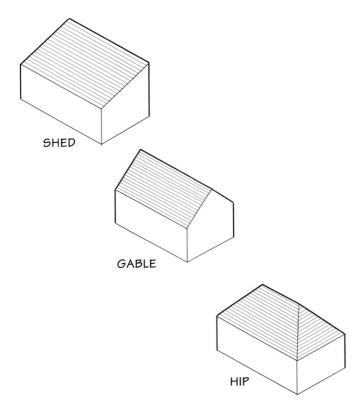

SHED

GABLE

HIP

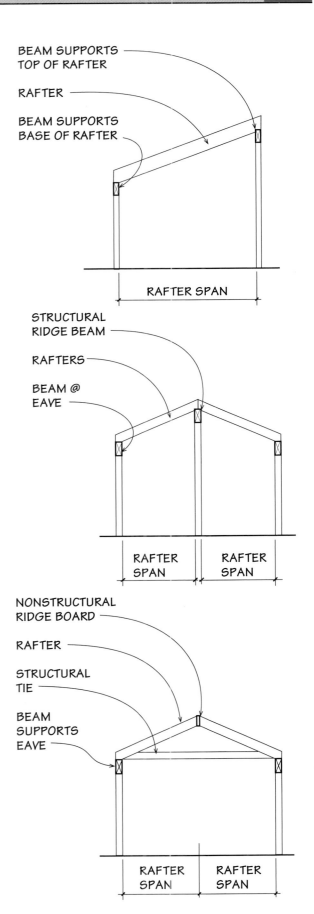

BEAM SUPPORTS TOP OF RAFTER

RAFTER

BEAM SUPPORTS BASE OF RAFTER

RAFTER SPAN

STRUCTURAL RIDGE BEAM

RAFTERS

BEAM @ EAVE

RAFTER SPAN   RAFTER SPAN

NONSTRUCTURAL RIDGE BOARD

RAFTER

STRUCTURAL TIE

BEAM SUPPORTS EAVE

RAFTER SPAN   RAFTER SPAN

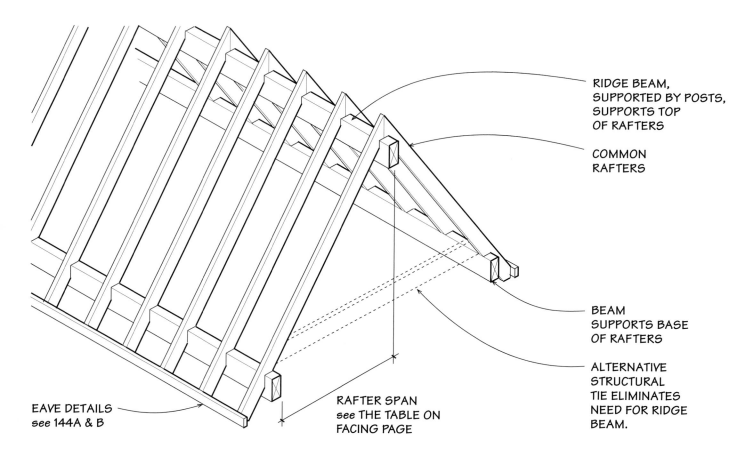

RIDGE BEAM, SUPPORTED BY POSTS, SUPPORTS TOP OF RAFTERS

COMMON RAFTERS

BEAM SUPPORTS BASE OF RAFTERS

ALTERNATIVE STRUCTURAL TIE ELIMINATES NEED FOR RIDGE BEAM.

EAVE DETAILS
see 144A & B

RAFTER SPAN
see THE TABLE ON FACING PAGE

 GABLE-ROOF FRAMING

A gable roof may be structured in one of two ways. First, a structural ridge beam may support the center of the roof, making the roof structurally like two shed roofs.

Alternatively, the ridge beam may be replaced with a nonstructural ridge board and ties that fasten at the base of pairs of opposing rafters. The ties resist the horizontal thrust resulting from the tendency of the rafter pairs to flatten out because they are not supported by a ridge beam at their top.

Rafter size, species, and grade	Rafter span (ft.)		
	12" O.C.	16" O.C.	24" O.C.
2x6 hem-fir #1	11.5	10.5	9.2
2x6 south. pine #1	12.0	10.9	9.6
2x6 Douglas-fir #1	12.2	11.1	9.7
2x8 hem-fir #1	15.2	13.8	12.1
2x8 south. pine #1	15.8	14.4	12.6
2x8 Douglas-fir #1	16.2	14.7	12.8
2x10 hem-fir #1	19.4	17.7	16.4
2x10 south. pine #1	20.3	18.4	17.1
2x10 Douglas-fir #1	20.6	18.7	17.4
2x12 hem-fir #1	23.6	21.5	18.7
2x12 south. pine #1	24.6	22.4	19.5
2x12 Douglas-fir #1	25.1	22.8	19.8

This table compares three species for a roof with a 30-psf live load. The table is for estimating purposes only.

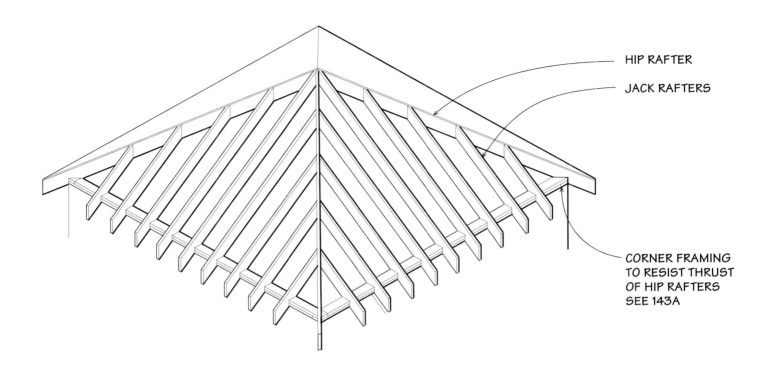

HIP RAFTER

JACK RAFTERS

CORNER FRAMING
TO RESIST THRUST
OF HIP RAFTERS
SEE 143A

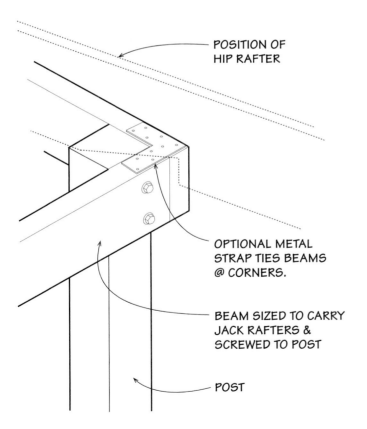

POSITION OF
HIP RAFTER

OPTIONAL METAL
STRAP TIES BEAMS
@ CORNERS.

BEAM SIZED TO CARRY
JACK RAFTERS &
SCREWED TO POST

POST

Hipped roofs can be made in the form of squares, rectangles, hexagons, octagons, and other regular shapes. A hipped roof has a structural assembly of hip rafters that lean against each other at their tops and are held in place at their bases by the continuity of the beams around the periphery of the roof. The beams that support a hipped roof must be solidly joined at their ends to resist the thrust of the hip rafters (see 143A).

(A) HIP-ROOF FRAMING

ROOFING AND MOISTURE PROTECTION—All roof structures are completed by the addition of sheathing that is selected to span between rafters. The sheathing and the rest of the roof structure is protected from the weather by the addition of flashing and roofing (see 144A & B). The flashing increases weather protection by creating an impervious layer and a drip edge—generally at the edge of the eave (sees 144C). The roofing, which can be made of a variety of materials, ranging from wood shingles to preformed metal panels, covers the entire roof surface to protect it from moisture. Care should be taken to choose roofing fasteners that do not protrude through exposed sheathing.

Roof drainage should be controlled to coordinate with other drainage systems (see chapter 1, p. 39-41).

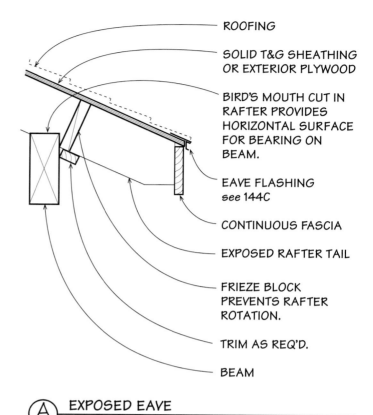

ROOFING

SOLID T&G SHEATHING OR EXTERIOR PLYWOOD

BIRD'S MOUTH CUT IN RAFTER PROVIDES HORIZONTAL SURFACE FOR BEARING ON BEAM.

EAVE FLASHING see 144C

CONTINUOUS FASCIA

EXPOSED RAFTER TAIL

FRIEZE BLOCK PREVENTS RAFTER ROTATION.

TRIM AS REQ'D.

BEAM

(A) **EXPOSED EAVE**

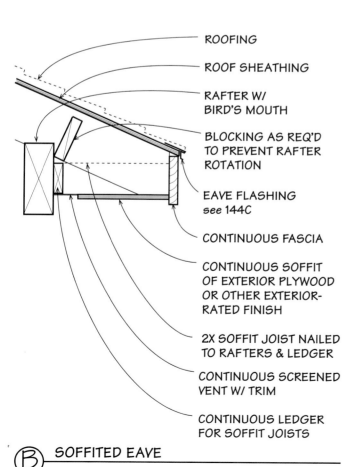

ROOFING

ROOF SHEATHING

RAFTER W/ BIRD'S MOUTH

BLOCKING AS REQ'D TO PREVENT RAFTER ROTATION

EAVE FLASHING see 144C

CONTINUOUS FASCIA

CONTINUOUS SOFFIT OF EXTERIOR PLYWOOD OR OTHER EXTERIOR-RATED FINISH

2X SOFFIT JOIST NAILED TO RAFTERS & LEDGER

CONTINUOUS SCREENED VENT W/ TRIM

CONTINUOUS LEDGER FOR SOFFIT JOISTS

(B) **SOFFITED EAVE**

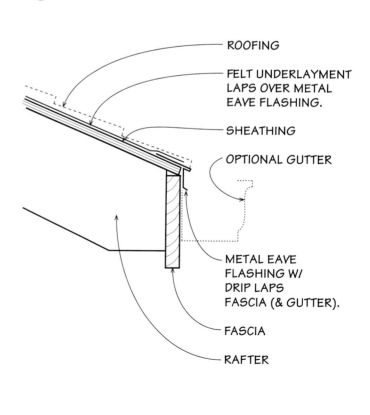

ROOFING

FELT UNDERLAYMENT LAPS OVER METAL EAVE FLASHING.

SHEATHING

OPTIONAL GUTTER

METAL EAVE FLASHING W/ DRIP LAPS FASCIA (& GUTTER).

FASCIA

RAFTER

(C) **EAVE FLASHING**

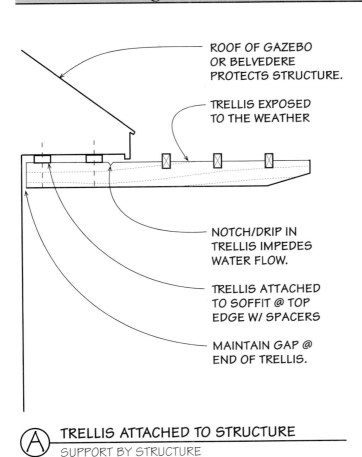

ROOF OF GAZEBO OR BELVEDERE PROTECTS STRUCTURE.

TRELLIS EXPOSED TO THE WEATHER

NOTCH/DRIP IN TRELLIS IMPEDES WATER FLOW.

TRELLIS ATTACHED TO SOFFIT @ TOP EDGE W/ SPACERS

MAINTAIN GAP @ END OF TRELLIS.

TRELLIS SUPPORTED BY POSTS

ROOF OF HOUSE REMAINS INTACT

NOTCH/DRIP IN BRACE IMPEDES WATER FLOW

BRACE ATTACHED TO SOFFIT @ TOP EDGE W/ SPACERS

MAINTAIN GAP @ END OF BRACE

(A) TRELLIS ATTACHED TO STRUCTURE
SUPPORT BY STRUCTURE

(B) TRELLIS ATTACHED TO STRUCTURE
SUPPORT BY POSTS

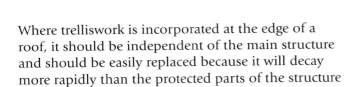

Where trelliswork is incorporated at the edge of a roof, it should be independent of the main structure and should be easily replaced because it will decay more rapidly than the protected parts of the structure (see 145A & B).

CHAPTER 5

BARBECUES AND OUTDOOR FIREPLACES

Outdoor living spaces have evolved over the years from basic decks and patios to full-fledged rooms. The kitchen and the family room have developed elaborate outdoor counterparts with increasingly sophisticated built environments that support comfortable and elegant outdoor living. The humble portable grill is giving way to expansive barbeques complete with cabinets, countertops, sinks, and even refrigerators, while the cozy comfort of the living-room fireplace has been translated for outdoor use.

Barbeques and outdoor fireplaces have many similar design concerns in that they both need a precise and carefully considered structure to control their potentially hazardous flames and comply with code requirements. They're also exposed to the weather and need to be durable enough to withstand all that Mother Nature can dish out. For these reasons, they are both generally made of masonry.

BARBECUES

The superb flavor of grilled foods, along with the pleasures of outdoor cooking, have combined to make barbecuing extremely popular in recent years. Most households now own some kind of portable grill and many are taking it a step further, creating built-in outdoor cooking amenities. From the simple fire pit to the complete outdoor kitchen, a well-planned and constructed outdoor cooking facility brings function and elegance to the backyard.

A complete site-built barbecue is like a kitchen cabinet installation in the great outdoors. In addition to the actual cooking unit, or grill, the barbecue may feature a generous countertop, storage cabinetry, a sink, even a refrigerator. The grill itself is typically manufactured (prefabricated) and may be fueled by gas (natural gas or propane) or charcoal. A wide range of manufactured units are available.

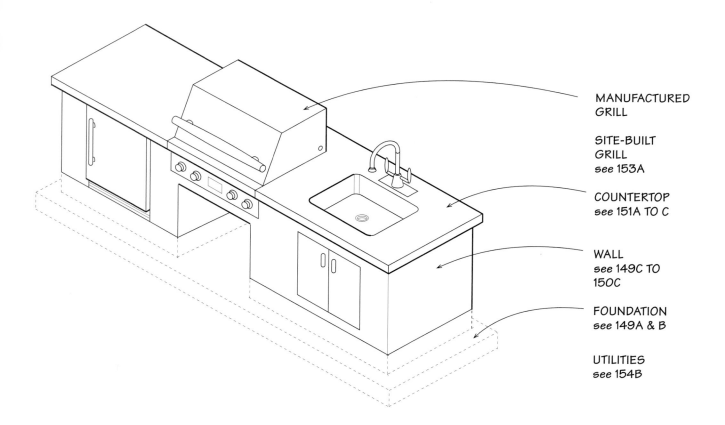

MANUFACTURED
GRILL

SITE-BUILT
GRILL
see 153A

COUNTERTOP
see 151A TO C

WALL
see 149C TO
150C

FOUNDATION
see 149A & B

UTILITIES
see 154B

When the grill is site-built as an integral part of the barbecue, however, it invariably uses charcoal as a fuel because of the difficulty of handcrafting a gas appliance. Site-built charcoal-fueled units can be as unique as the imagination of the designer.

BUILDING CODES—Site-built barbecues are usually regulated by building codes because they generally incorporate utilities such as electricity, natural gas, or plumbing. Electrical power is frequently required for the pilot light in a manufactured gas grill and is often desired for convenience, for lighting, or for a refrigerator or other appliance.

Plumbing of both supply and wastewater is regulated by code. Plumbing codes also govern the design and installation of natural gas lines. The gas grill industry recommends a 24-in. clearance between their products and any combustible materials. Check local building codes and manufacturer's recommendations before designing a barbecue.

THE BARBECUE CABINET—The basic form of most barbecues resembles kitchen cabinetry. Based on human proportions, it provides a level surface upon which to prepare and serve food, and incorporates a grill upon which the cooking is actually done. Storage and other amenities can be as simple or extravagant as the client's resources and imagination allow.

The basic components of this outdoor cabinetry include a foundation, cabinet walls, and a countertop. Because of weather exposure and heat and flames from the grill, it is best if the basic materials for such construction are both noncombustible and weather resistant. This limits the practical construction choices to masonry or concrete with tile or metal highlights.

Reinforcement of masonry (or concrete) is advisable in earthquake zones or locations where the soil is unstable. Otherwise, because these structures are not tall, reinforcement is not usually necessary for the foundation or walls. When required, reinforcement is typically done with #3 or #4 rebar embedded in the masonry or concrete.

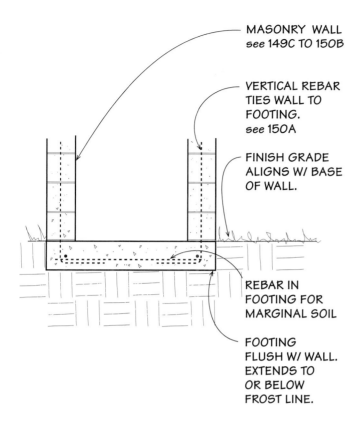

MASONRY WALL
see 149C TO 150B

VERTICAL REBAR
TIES WALL TO
FOOTING.
see 150A

FINISH GRADE
ALIGNS W/ BASE
OF WALL.

REBAR IN
FOOTING FOR
MARGINAL SOIL

FOOTING
FLUSH W/ WALL.
EXTENDS TO
OR BELOW
FROST LINE.

Ⓐ FLUSH FOOTING

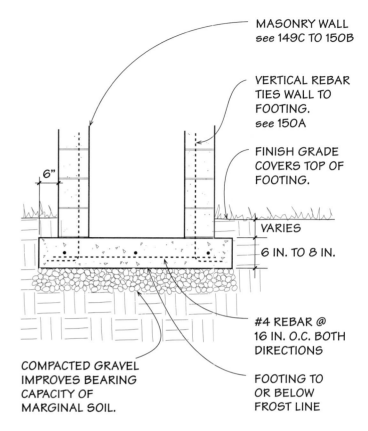

MASONRY WALL
see 149C TO 150B

VERTICAL REBAR
TIES WALL TO
FOOTING.
see 150A

FINISH GRADE
COVERS TOP OF
FOOTING.

VARIES

6 IN. TO 8 IN.

#4 REBAR @
16 IN. O.C. BOTH
DIRECTIONS

FOOTING TO
OR BELOW
FROST LINE

COMPACTED GRAVEL
IMPROVES BEARING
CAPACITY OF
MARGINAL SOIL.

6"

Ⓑ SPREAD FOOTING

The foundation—The foundation for a built-in barbecue should be designed to match the weight of the structure to the properties of the soil that will support it. Most soils will support 1,500 lb. per sq. ft. (psf), which is the equivalent of a solid concrete mass 10 ft. tall. Therefore, most 3-ft. masonry barbecue structures can be supported effectively on a simple concrete footing no wider than the structure itself (see 149A).

However, to be conservative, it is recommended that the footing be 8 in. thick and extend 6 in. beyond the structure (see 149B). Footings should be constructed on undisturbed soil that is free of organic material, and expansive clay soils should be avoided. The footing base should extend below the frost line. Consult a local engineer if there are any questions about the ability of the soil to support such a structure.

The walls—Masonry, whether stone, brick, or concrete masonry, is the construction type of choice for the walls that support the grill and countertop (see 149C). Masonry is durable, weather-resistant, non-

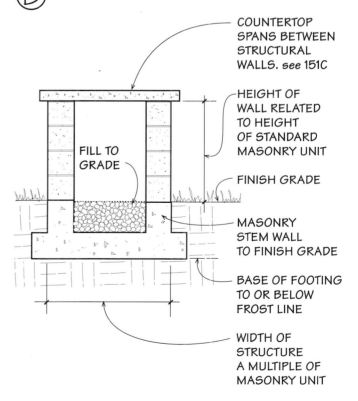

COUNTERTOP
SPANS BETWEEN
STRUCTURAL
WALLS. see 151C

HEIGHT OF
WALL RELATED
TO HEIGHT
OF STANDARD
MASONRY UNIT

FINISH GRADE

MASONRY
STEM WALL
TO FINISH GRADE

BASE OF FOOTING
TO OR BELOW
FROST LINE

WIDTH OF
STRUCTURE
A MULTIPLE OF
MASONRY UNIT

FILL TO
GRADE

Ⓒ BASIC STRUCTURE
SECTION

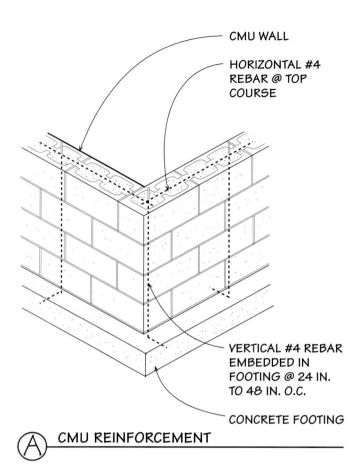

CMU WALL

HORIZONTAL #4 REBAR @ TOP COURSE

VERTICAL #4 REBAR EMBEDDED IN FOOTING @ 24 IN. TO 48 IN. O.C.

CONCRETE FOOTING

Ⓐ CMU REINFORCEMENT

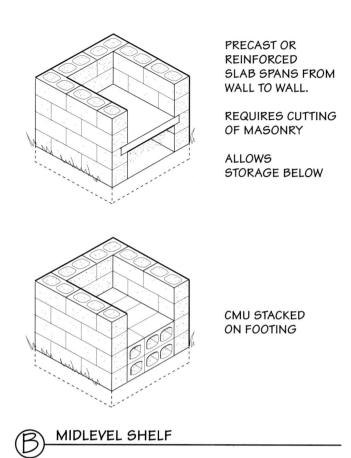

PRECAST OR REINFORCED SLAB SPANS FROM WALL TO WALL.

REQUIRES CUTTING OF MASONRY

ALLOWS STORAGE BELOW

CMU STACKED ON FOOTING

Ⓑ MIDLEVEL SHELF

combustible, and relatively easy to build. Concrete masonry units (CMUs) are typically used because they are easily laid, inexpensive, and easily reinforced (see 150A). Brick is also a good option. Concrete construction also may be used, but it must be formed—an expensive proposition for a typically small and custom structure.

Intermediate-level shelves within the structure support the grill (whether site-built or manufactured) or can be used for storage. These shelves are typically made of masonry or concrete (precast or site-cast) that is simply integrated with the masonry walls (see 150B).

Light-gauge steel framing is an alternative to masonry or concrete for the cabinet walls (see 150C). The framing can be stabilized with the application of cement backer-board to which a variety of veneers may be applied. In the long run, however, masonry will outlast steel in an exposed environment.

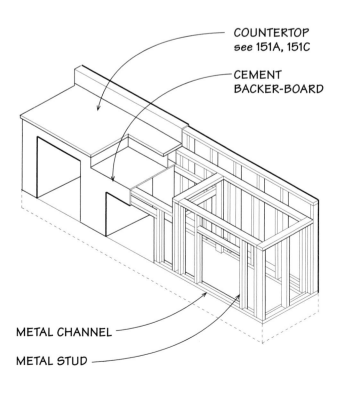

COUNTERTOP
see 151A, 151C

CEMENT BACKER-BOARD

METAL CHANNEL

METAL STUD

Ⓒ LIGHT-GAUGE STEEL WALLS

The countertop—Countertops should be noncombustible, weather resistant, and easily cleaned. Therefore, most barbecue countertops are made of tile (ceramic or stone) adhered to a concrete base. The concrete spans the distance between the walls of the cabinet and provides a solid base for the tiles (see 151A & B). The tiles provide pattern, color, and a hard and easily cleaned surface.

In warm, dry climates where moisture isn't much of an issue, tiles are often applied to a preservative-treated plywood backing that takes the place of the concrete base (see 151C). Sheet metal, such as stainless steel or copper, applied over a wood base also can make a beautiful and durable countertop. However, caution should be used to protect any structural wood pieces from the heat of the cooking appliances.

Concrete also can be troweled smooth, sealed, and left as a finished surface. It is softer and more porous than most tiles, but it can be more durable in the long run if treated regularly with sealer to guard against moisture and stains.

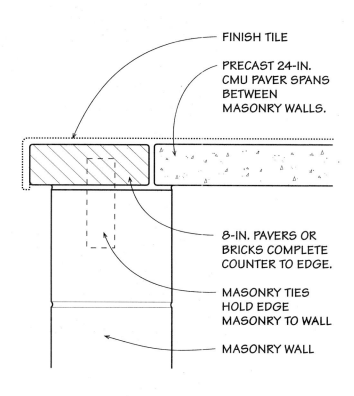

FINISH TILE

PRECAST 24-IN. CMU PAVER SPANS BETWEEN MASONRY WALLS.

8-IN. PAVERS OR BRICKS COMPLETE COUNTER TO EDGE.

MASONRY TIES HOLD EDGE MASONRY TO WALL

MASONRY WALL

Ⓑ PRECAST COUNTERTOP

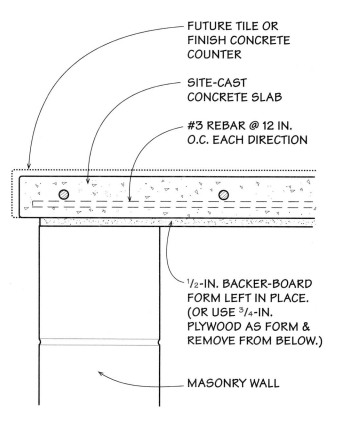

FUTURE TILE OR FINISH CONCRETE COUNTER

SITE-CAST CONCRETE SLAB

#3 REBAR @ 12 IN. O.C. EACH DIRECTION

1/2-IN. BACKER-BOARD FORM LEFT IN PLACE. (OR USE 3/4-IN. PLYWOOD AS FORM & REMOVE FROM BELOW.)

MASONRY WALL

Ⓐ SITE-CAST COUNTERTOP

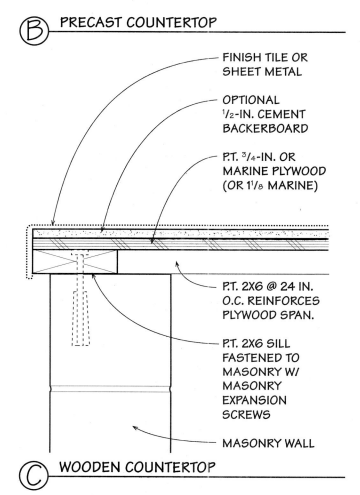

FINISH TILE OR SHEET METAL

OPTIONAL 1/2-IN. CEMENT BACKERBOARD

P.T. 3/4-IN. OR MARINE PLYWOOD (OR 1 1/8 MARINE)

P.T. 2X6 @ 24 IN. O.C. REINFORCES PLYWOOD SPAN.

P.T. 2X6 SILL FASTENED TO MASONRY W/ MASONRY EXPANSION SCREWS

MASONRY WALL

Ⓒ WOODEN COUNTERTOP

Wall finishes—The masonry (or metal framing) that forms the walls of a barbecue cabinet can be finished in a variety of ways. Masonry walls may be left unfinished, and the range of colors and textures available in both CMUs and brick makes this approach popular.

Masonry, concrete, or cement backer-board also may be stuccoed, tiled, or covered with a masonry veneer (see 152A) or it can be coated with a masonry paint or opaque sealer. Finishes should be designed to accommodate cabinet doors and other exposed details carefully (see 152B).

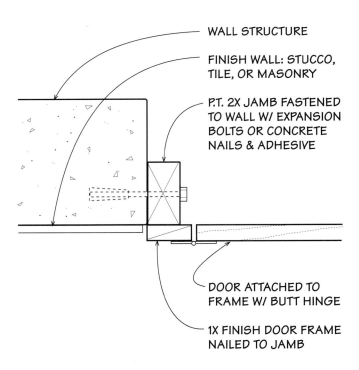

WALL STRUCTURE

FINISH WALL: STUCCO, TILE, OR MASONRY

P.T. 2X JAMB FASTENED TO WALL W/ EXPANSION BOLTS OR CONCRETE NAILS & ADHESIVE

DOOR ATTACHED TO FRAME W/ BUTT HINGE

1X FINISH DOOR FRAME NAILED TO JAMB

NOTE: MANY OTHER DETAILS WORK IN THIS SITUATION. THIS IS ONE OF THE SIMPLEST & MOST COMMON.

Ⓑ **CABINET DOOR DETAIL**
PLAN VIEW

THE GRILL—When it comes to the grill itself, there are two basic approaches—buy a manufactured unit and insert it into the barbecue cabinet, or design and build a grill from scratch. Manufactured insert units are much more expensive but provide more control over cooking, while site-built grills usually take much more time to construct and are limited to being charcoal fueled. Generally, the manufactured unit is a cooktop that is integrated into the barbecue cabinet. It is also possible, however, to buy a freestanding grill and build a cabinet around it as a kind of garage.

Manufactured grills—There are myriad manufactured barbecue grills available that are designed to be inserted into a site-built cabinet. The most common are essentially the same as portable grills but without the cart and wheels. Fueled by either gas or charcoal, the best units are made of coated steel or stainless, both of which are more durable than painted aluminum. Smoker trays, which hold wood chips to impart a smoked flavor to foods, have become a popular option recently.

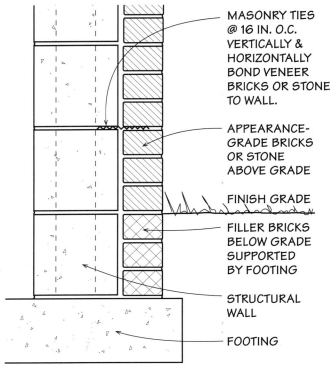

MASONRY TIES @ 16 IN. O.C. VERTICALLY & HORIZONTALLY BOND VENEER BRICKS OR STONE TO WALL.

APPEARANCE-GRADE BRICKS OR STONE ABOVE GRADE

FINISH GRADE

FILLER BRICKS BELOW GRADE SUPPORTED BY FOOTING

STRUCTURAL WALL

FOOTING

Ⓐ **MASONRY VENEER**
SECTION

Gas-fueled units are invariably self-contained and include a firebox with controls, a cooking grate, and a retractable cover. They require a gas supply (propane or natural gas) and an electrical connection for the pilot ignition and optional electrical conveniences. Natural gas is more trouble-free than propane in the long run because there are no tanks to be refilled, but natural gas is more difficult to install because the grill must be connected to a permanent gas supply.

Charcoal-fueled units are also self-contained and typically include a firebox with adjustable air flow, an adjustable cooking grate, and a vented retractable lid. The best units have convenient fire-control and ash-removal systems. Unlike the gas-fueled units, charcoal grills require no utilities unless they are equipped with options such as an electrical rotisserie.

Site-built grills—Site-built grills, which have no manufactured parts except an adjustable cooking grate and a rack below to hold the fuel, are inexpensive and can be made in virtually any size or shape (see 153A).

The cooking grate and fuel rack may be purchased off-the-shelf from a fireplace shop, gourmet cooking supply, or home-improvement store; or they may be custom-made. More elaborate adjustable-height cooking grates that are designed to be installed over a site-built firebox are also available.

UTILITIES—Gas, electricity, and water are often required for a built-in home barbecue. All of these utilities may be piped underground and should be carefully planned before footings are poured. They all may be regulated by code, as well, and local codes should be checked during the design process.

Gas—Gas for burners must be supplied in approved piping that's buried at an approved depth. The pipe terminates at a shutoff valve with a pressure regulator located below the gas-burning appliance and is connected to the appliance with a flexible connection. In locations where underground piping is difficult or natural gas is unavailable, a propane tank may be located in the barbecue structure as long as it is not fully enclosed.

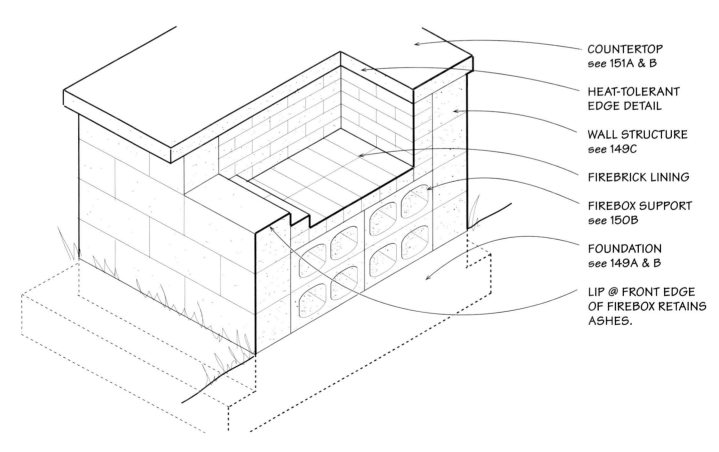

COUNTERTOP
see 151A & B

HEAT-TOLERANT
EDGE DETAIL

WALL STRUCTURE
see 149C

FIREBRICK LINING

FIREBOX SUPPORT
see 150B

FOUNDATION
see 149A & B

LIP @ FRONT EDGE
OF FIREBOX RETAINS
ASHES.

(A) SITE-BUILT GRILL
SECTION @ FIREBOX

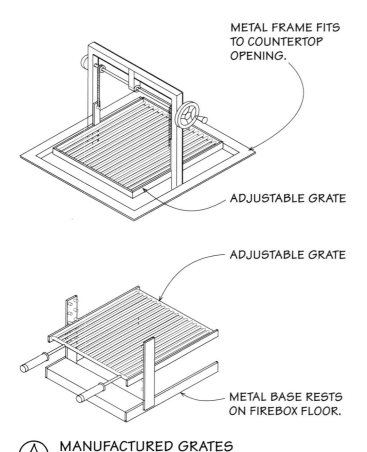

METAL FRAME FITS TO COUNTERTOP OPENING.

ADJUSTABLE GRATE

ADJUSTABLE GRATE

METAL BASE RESTS ON FIREBOX FLOOR.

(A) MANUFACTURED GRATES

Electricity—Line-voltage electricity is usually required at a barbecue for electronic ignition of burners, convenience receptacles, line-voltage lighting, or refrigerator and any other appliances. The power line is typically buried and enters the barbecue structure from below (see Chapter. 2, pp. 168–169). Once inside the grill, all wiring must be enclosed in conduit, and care should be taken to route it away from the heat of the grill.

Water—Running water is a real convenience at a barbecue, so a sink is often installed in the countertop. Cold water can be piped underground, but in cold climates, the supply piping should be drainable in winter months to prevent freezing.

Hot water may be piped in insulated pipes from the house to the barbecue. Alternatively, a small, demand water heater can be located near the outdoor sink. A drain from the sink must be vented and routed to the building sewer in some states, but in others, the drain may be run to the surface of the ground.

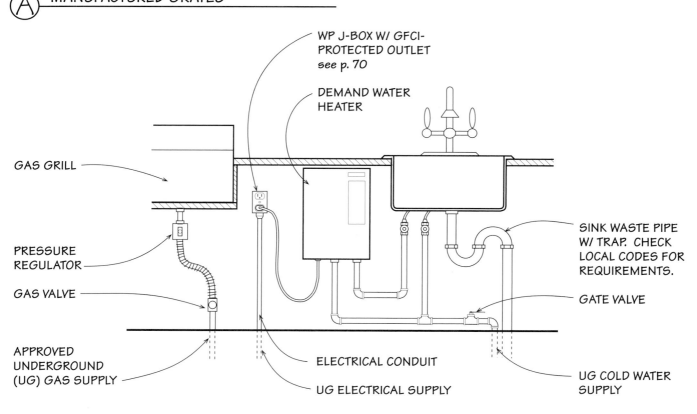

WP J-BOX W/ GFCI-PROTECTED OUTLET see p. 70

DEMAND WATER HEATER

GAS GRILL

PRESSURE REGULATOR

GAS VALVE

APPROVED UNDERGROUND (UG) GAS SUPPLY

ELECTRICAL CONDUIT

UG ELECTRICAL SUPPLY

SINK WASTE PIPE W/ TRAP. CHECK LOCAL CODES FOR REQUIREMENTS.

GATE VALVE

UG COLD WATER SUPPLY

(B) UTILITIES

FIREPLACES

Outdoor fireplaces and fire pits are becoming increasingly prevalent. A fire's warmth and mesmerizing flames, along with its pleasant sounds and smells, strike a resonant chord in all of us who share thousands of years of evolution beside the open fire. In fact, fire is such a compelling social focus that outdoor fireplaces are cropping up even in the warmest climatic zones, where they generally serve no practical function. While fireplaces and fire pits are designed for comfort and human interaction, safety and environmental responsibility also should be major considerations.

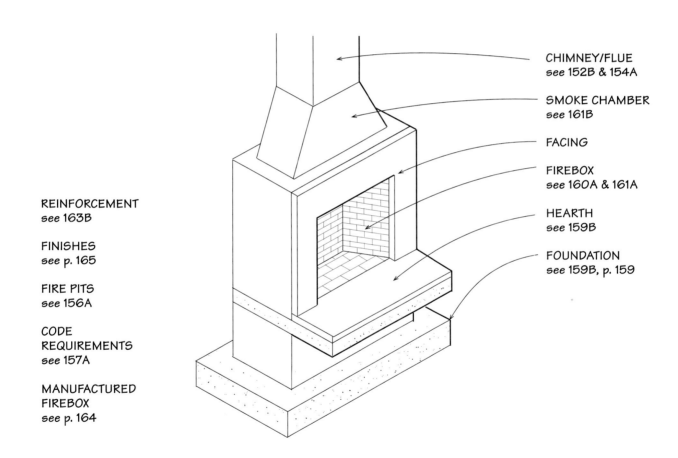

REINFORCEMENT
see 163B

FINISHES
see p. 165

FIRE PITS
see 156A

CODE
REQUIREMENTS
see 157A

MANUFACTURED
FIREBOX
see p. 164

CHIMNEY/FLUE
see 152B & 154A

SMOKE CHAMBER
see 161B

FACING

FIREBOX
see 160A & 161A

HEARTH
see 159B

FOUNDATION
see 159B, p. 159

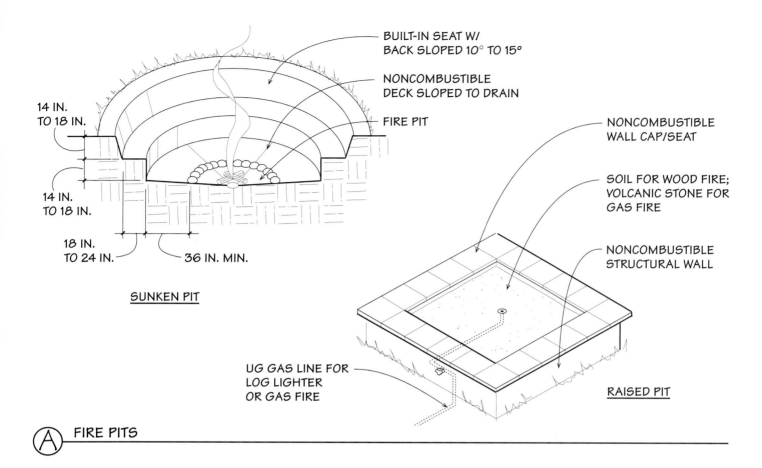

14 IN.
TO 18 IN.

14 IN.
TO 18 IN.

18 IN.
TO 24 IN.

36 IN. MIN.

BUILT-IN SEAT W/
BACK SLOPED 10° TO 15°

NONCOMBUSTIBLE
DECK SLOPED TO DRAIN

FIRE PIT

SUNKEN PIT

NONCOMBUSTIBLE
WALL CAP/SEAT

SOIL FOR WOOD FIRE;
VOLCANIC STONE FOR
GAS FIRE

NONCOMBUSTIBLE
STRUCTURAL WALL

UG GAS LINE FOR
LOG LIGHTER
OR GAS FIRE

RAISED PIT

(A) FIRE PITS

Fire pits are designated places in the garden in which it is safe to build an open fire. The simplest fire pit is merely a depression in the ground in an area that has been cleared of combustible materials and surrounded with comfortable seating—much like a fire pit in a campground. The most elaborate pits are surrounded with masonry walls, integrated seating, electric lighting, and a gas lighter (see 156A). Because fire pits have such basic design requirements, this chapter will focus on fireplaces.

Outdoor fireplaces have evolved from fire pits, influenced by their indoor cousins originally used for heating and cooking. Today the indoor fireplace princi-

pally serves to create a warm atmosphere or mood. However, the outdoor fireplace is still used for heating and cooking. As a source of heat, the outdoor fireplace extends the use of outdoor spaces longer into the night and deeper into the transitional seasons of spring and fall. Because the heat radiated by a fireplace is directional (not radial like a fire pit), careful planning is important to locate it in a position that allows natural social groups to take advantage of it.

BUILDING AND ZONING CODES–The regulation of outdoor fire-pit and fireplace construction varies considerably from jurisdiction to jurisdiction. In some areas, they are not allowed at all because of strict air pollution or fire-safety zoning regulations. In other areas, they are allowed but can only be used on days determined to have favorable atmospheric conditions.

Where allowed, fire pits are not regulated by the building code (unless they have a gas lighter or gas flame), but outdoor fireplaces are governed by the code when over 10 ft. tall (which they usually are). The building code specifies footing size, masonry thickness, reinforcement, the proportions of firebox, hearth, and flue, and minimum clearance to combustible materials (see 157A). The code is written for fireplaces incorporated within a structure, however, so many of the regulations do not apply to outdoor fireplaces. Check with your local building department to determine the extent of codes in your area.

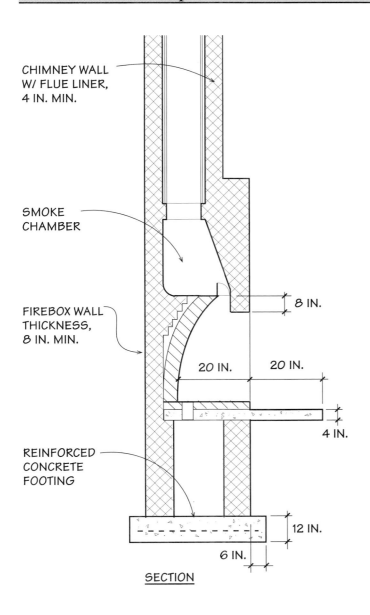

CHIMNEY WALL W/ FLUE LINER, 4 IN. MIN.

SMOKE CHAMBER

FIREBOX WALL THICKNESS, 8 IN. MIN.

REINFORCED CONCRETE FOOTING

8 IN.

20 IN.   20 IN.

4 IN.

12 IN.

6 IN.

SECTION

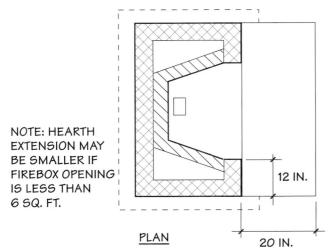

NOTE: HEARTH EXTENSION MAY BE SMALLER IF FIREBOX OPENING IS LESS THAN 6 SQ. FT.

12 IN.

PLAN

20 IN.

(A) **CODE REQUIREMENTS**

MASONRY FIREPLACE

DESIGN CONSIDERATIONS—While similar in many ways to indoor fireplaces, it is a mistake to apply indoor design to outdoor units without careful consideration of differences. A comparison of the physical characteristics of indoor and outdoor fireplaces reveals the unique qualities of each.

- Creating a stable outdoor fireplace is a challenge because it is a generally tall and massive element not braced by an attached building. An outdoor fireplace usually requires a wider footing and more reinforcement than an interior fireplace. Alternatively, an outdoor fireplace can be tied to the structure of the house.

- Because an outdoor fireplace is subjected to wind, it is more difficult to control smoking from the firebox. However, a carefully designed flue can eliminate most of the smoke.

- The shape of an outdoor firebox should place a higher priority on heat radiation than its indoor counterpart, which generally focuses on minimizing the amount of smoke leaking into a room. The Rumford style firebox, which was created in the 1790s and reintroduced into domestic architecture about 25 years ago, is ideal for exterior use where heat is important (see 159A).

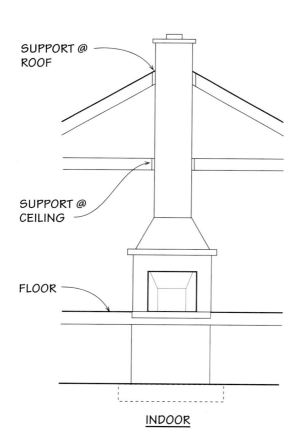

INDOOR

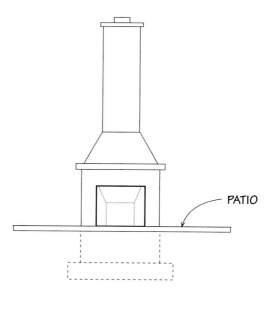

OUTDOOR

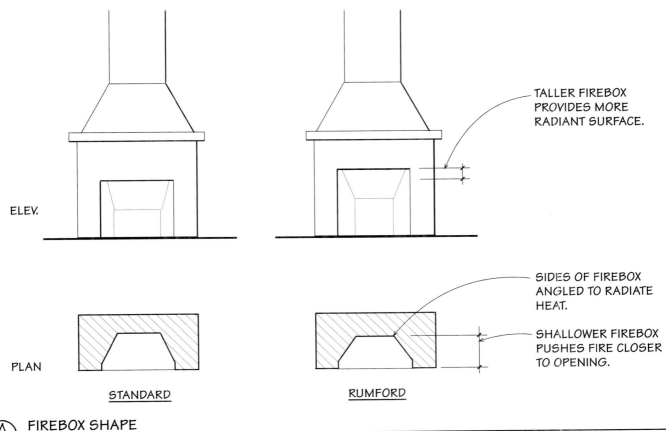

ELEV.

TALLER FIREBOX PROVIDES MORE RADIANT SURFACE.

SIDES OF FIREBOX ANGLED TO RADIATE HEAT.

SHALLOWER FIREBOX PUSHES FIRE CLOSER TO OPENING.

PLAN

STANDARD

RUMFORD

(A) FIREBOX SHAPE

STRUCTURE AND COMPONENTS—The parts of a site-built masonry fireplace are constructed by stacking one masonry unit on another—just as a masonry wall is built one course at a time. A masonry foundation wall, a hearth, a firebox, a smoke chamber, and a flue are all stacked on a concrete footing. In earthquake and high-wind zones, these components are tied together with vertical and horizontal rebar.

Foundation—The residential building code requires that an indoor masonry fireplace have a footing that is 12 in. deep and extends 6 in. beyond the masonry in all directions (see 159B). In most cases, these specifications also will be adequate for the vertical loads of an outdoor masonry fireplace.

However, the ground under freestanding outdoor fireplaces is more prone to moisture saturation, which can weaken the foundation. A freestanding fireplace is also inherently weaker than an attached one

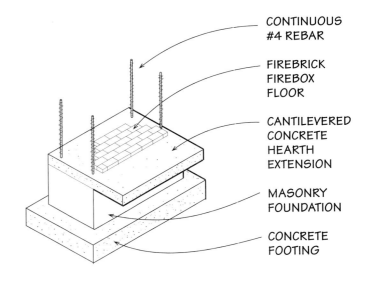

CONTINUOUS #4 REBAR

FIREBRICK FIREBOX FLOOR

CANTILEVERED CONCRETE HEARTH EXTENSION

MASONRY FOUNDATION

CONCRETE FOOTING

(B) **MASONRY FIREPLACE**
FOUNDATION TO HEARTH

because it doesn't have the additional support of another structure. For both of these reasons, it is recommended that freestanding fireplaces have wider footings than are required by code. It usually is wise to consult a structural engineer to size the footing and design its reinforcing.

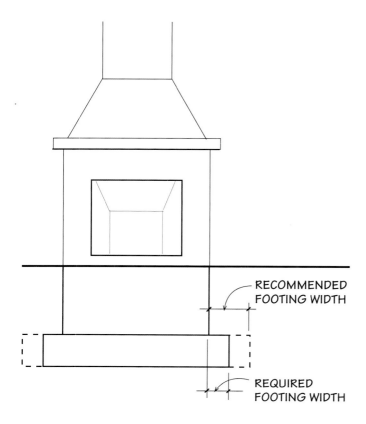

RECOMMENDED FOOTING WIDTH

REQUIRED FOOTING WIDTH

Hearth extension—The hearth extension in front of the firebox provides protection from burning embers that may escape from the fire (see 159B). The hearth extension, therefore, must be made of noncombustible materials. If the area in front of the firebox is a paved terrace, there is no need for a separate hearth extension. If, however, the area in front of the firebox is a wooden deck, then a separate, noncombustible hearth extension must be constructed.

A hearth extension may be cantilevered from the masonry fireplace mass or it may be supported on an extended footing. In any case, it may not be supported by combustible material.

Firebox—The firebox and the smoke chamber above it are the only truly complex parts of a masonry fireplace. The angles between firebox walls, the need to span the large firebox opening, and the sloping faces of the throat leading up to the smoke chamber all contribute to the complexity, compelling most homeowners to leave the job to a skilled mason.

The firebox floor and walls should be made with refractory firebrick, which is designed to withstand the heat of a wood fire (see 160A & 161A). These special bricks are laid with very thin fireclay joints up to the full height of the firebox, including the throat zone at the top where a transition is made to the smoke chamber.

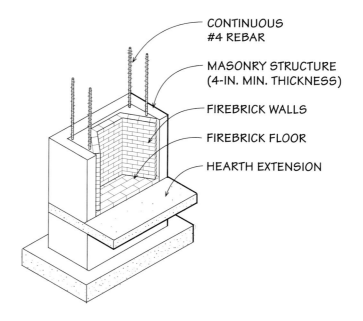

CONTINUOUS #4 REBAR

MASONRY STRUCTURE (4-IN. MIN. THICKNESS)

FIREBRICK WALLS

FIREBRICK FLOOR

HEARTH EXTENSION

Ⓐ MASONRY FIREPLACE
TO TOP OF FIREBOX OPENING

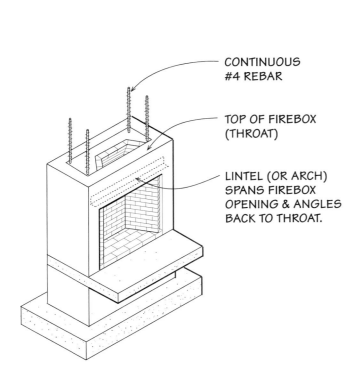

CONTINUOUS #4 REBAR

TOP OF FIREBOX (THROAT)

LINTEL (OR ARCH) SPANS FIREBOX OPENING & ANGLES BACK TO THROAT.

Ⓐ MASONRY FIREPLACE
TO THROAT

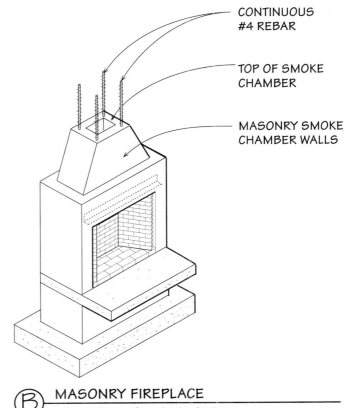

CONTINUOUS #4 REBAR

TOP OF SMOKE CHAMBER

MASONRY SMOKE CHAMBER WALLS

Ⓑ MASONRY FIREPLACE
TO TOP OF SMOKE CHAMBER

Smoke Chamber—The smoke chamber is the transitional area between firebox and flue. The smoke chamber, in conjunction with the flue, promotes the upward flow of smoke. Its shape is designed to turn falling eddies of smoke within the flue and redirect them upward. The proportions of the smoke chamber have been developed over hundreds of years, and, like the firebox, are usually best left to an experienced mason (see 161B).

Most smoke chambers are shaped with regular brick, and then parged (covered with cement plaster) to form a smooth surface inside the chamber. Smoke chambers also may be made with terra-cotta components preformed to the shape of the chamber.

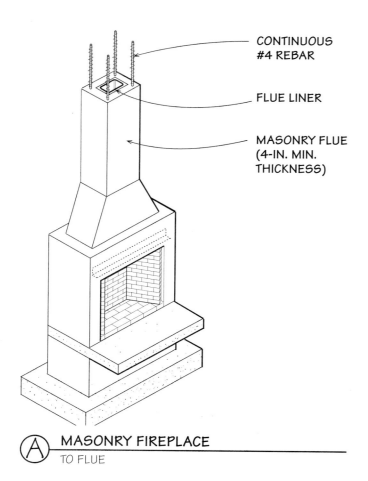

CONTINUOUS #4 REBAR

FLUE LINER

MASONRY FLUE (4-IN. MIN. THICKNESS)

Ⓐ MASONRY FIREPLACE
TO FLUE

Flue—Along with the smoke chamber, the flue is an important element in getting the fireplace to draw. The principle variables in flue design are its cross-sectional area and its height. There is a range of recommended flue sizes for fireplaces, but for outdoor units, bigger tends to be better. A cross-sectional area approximately one-tenth of the area of the firebox opening works well. For example, a firebox with a 40-in. by 40-in. (1,600 sq. in.) opening should have a flue with a cross-sectional area that's approximately 160 sq. in.

The height of the flue should be at least three to four times the height of the firebox opening and should extend 2 ft. to 4 ft. above any roof within 10 ft.

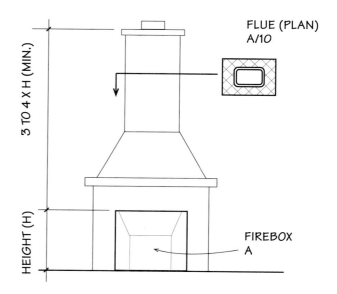

FLUE (PLAN)
A/10

3 TO 4 X H (MIN.)

HEIGHT (H)

FIREBOX
A

The best flues are made of masonry with a fireclay liner (see 163A). The fireclay is smooth, which promotes the free flow of smoke, and it is dense, which helps retain heat—further warming the smoke to make it lighter and more likely to rise. Lighter, rough-textured and dense pumice liners are also available. To increase smoothness and strength, the flue may be parged (plastered) with cement plaster.

Reinforcement—Masonry fireboxes and flues should be reinforced both vertically and horizontally to resist the lateral forces of earthquakes and high winds. Typically, a #4 rebar run continuously at each corner of the construction is adequate for vertical reinforcement. The vertical rebars should be tied at the base to vertical rebars that are structurally embedded in the footing (see 163B).

Horizontal reinforcement at 48-in. intervals will tie the vertical bars together. With CMU construction, a #4 rebar bond beam works well, while joint reinforcement is adequate for smaller masonry units (see pp. 93-95).

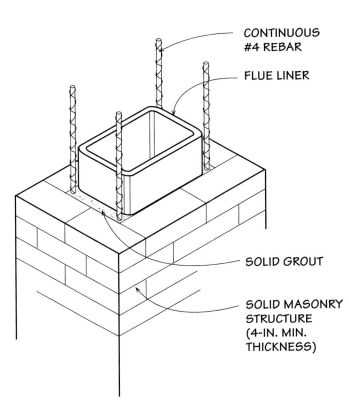

CONTINUOUS #4 REBAR

FLUE LINER

SOLID GROUT

SOLID MASONRY STRUCTURE (4-IN. MIN. THICKNESS)

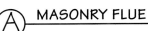 **MASONRY FLUE**

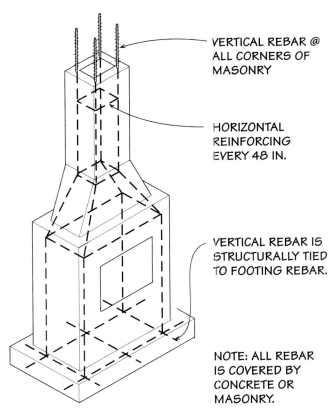

VERTICAL REBAR @ ALL CORNERS OF MASONRY

HORIZONTAL REINFORCING EVERY 48 IN.

VERTICAL REBAR IS STRUCTURALLY TIED TO FOOTING REBAR.

NOTE: ALL REBAR IS COVERED BY CONCRETE OR MASONRY.

 FIREPLACE REINFORCEMENT

MANUFACTURED FIREPLACES—Most outdoor fireplaces are built entirely on the site out of masonry. However, it is possible to eliminate much of the on-site work by using a manufactured firebox and flue. The manufactured firebox is a metal enclosure including a fully operational firebox lined with refractory materials and containing a smoke chamber. The flue is also made of metal and consists of interlocking sections that stack on top of the firebox. Most manufactured outdoor fireboxes and flues are made of stainless steel.

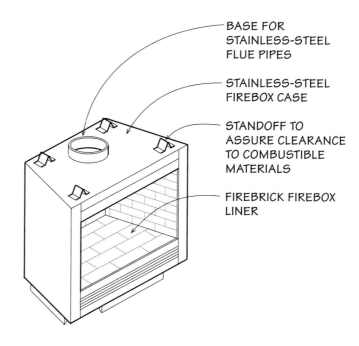

BASE FOR STAINLESS-STEEL FLUE PIPES

STAINLESS-STEEL FIREBOX CASE

STANDOFF TO ASSURE CLEARANCE TO COMBUSTIBLE MATERIALS

FIREBRICK FIREBOX LINER

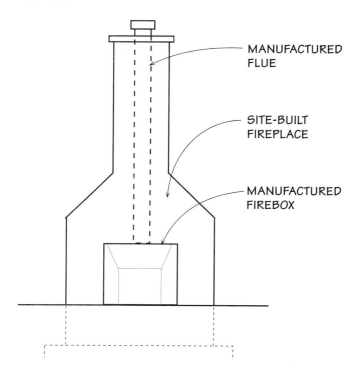

MANUFACTURED FLUE

SITE-BUILT FIREPLACE

MANUFACTURED FIREBOX

(A) MANUFACTURED FIREBOX

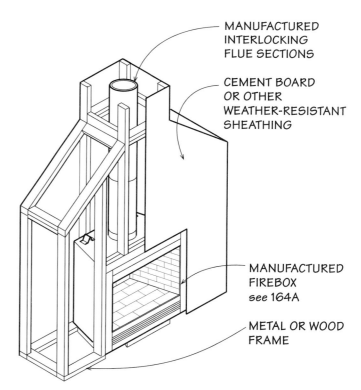

MANUFACTURED INTERLOCKING FLUE SECTIONS

CEMENT BOARD OR OTHER WEATHER-RESISTANT SHEATHING

MANUFACTURED FIREBOX
see 164A

METAL OR WOOD FRAME

NOTE: FOLLOW MANUFACTURER'S SPECIFICATIONS FOR FIREBOX & FLUE CLEARANCES.

(B) FRAMED FIREPLACE

The use of a manufactured firebox avoids the complexities of masonry firebox design and construction, but does require the construction of an enclosure for the manufactured components. The enclosure also must provide lateral stability for the segmented flue. A wood or light-gauge steel framed structure is often used for this purpose because this allows masonry construction to be avoided altogether (see 164B). The framed structure is sheathed in cement backerboard panels that are then finished with stucco, tile, synthetic stone, or some other relatively lightweight material. Alternatively, the enclosure may be constructed with the same materials as would be used for a masonry firebox.

There may be an economic advantage to the use of a manufactured firebox and flue in some cases—especially where the price of masonry construction is high. However, these structures are not likely to endure nearly as long as masonry, and the finishing options are limited when light framing is used for the enclosure. Furthermore, the size and shape of the firebox is severely limited because there are very few models from which to choose.

FIREPLACE FINISH MATERIALS—Outdoor fireplaces may be finished with any noncombustible material suitable for outdoor use. Stucco, brick, stone, and synthetic stone are common. Some materials (stucco, synthetic stone) must be applied over a structural backup material such as CMU or backerboard. Other materials (brick and stone) may be applied as a veneer over CMU or may be used for structure themselves, eliminating the need for CMU backup. The wide range of options presents a great opportunity to coordinate the fireplace finish with retaining walls, paving, and other landscape features.

GAS LIGHTERS AND FLAMES—Gas log lighters are available for outdoor use, and are used in fire pits as well as fireplaces. These contrivances simplify the task of lighting a fire by eliminating the need for kindling and the gradual building of a full-size flame. They do add long-term maintenance to the equation, however, because they are made of parts that can corrode or wear out.

Gas flames also can be incorporated into a fire pit or fireplace. These devices altogether eliminate the need for wood as a fuel and do away with smoke as well, but the fire is reduced to a relatively constant flame that emerges through static ceramic logs or lava rocks.

The gas for lighters or flames is usually piped underground to the fireplace or fire pit. Local codes should be consulted before installing any of these devices.

CHAPTER 6

WATER IN THE LANDSCAPE

Water can take on many different forms and perform many different functions in a yard or garden. Water can be a visual element, an auditory element, a functional element (such as a storm-water collection pond), or an element for play, relaxation, exercise, and therapy.

Water features can be naturalistic or formal and geometric. They can be symmetrical or asymmetrical, in-ground or raised (see 167A). Some may house plants, fish, or amphibians, while others are built for people. In many cases, a water feature takes on several roles at once, providing the homeowner with an invaluable resource that provides a great deal of pleasure.

However, there are many issues to address when considering a water feature that goes beyond initial installation. Most water features require a level of maintenance that can become overwhelming (in terms of time, energy, and money) unless care is taken during design and construction. Safety is another concern, especially with small children; even a few inches of water can cause a potentially fatal accident.

Check local codes and zoning laws regarding fencing, permits, and other water-feature regulations, such as vicinity to a property line or a utility easement. Also investigate how close out-of-water equipment can be to the water itself, and how and where the GFCI needs to be located.

PONDS, WATERFALLS, FOUNTAINS, AND STREAMS

While generally not as large or complex as a swimming pool, ponds, waterfalls, fountains, and streams still require much forethought to ensure proper design and construction. They can be expensive and time-consuming to maintain if not installed properly, and they can take up substantial space in the landscape, which may interfere with a site's maximum usefulness.

So, before a feature is decided on, it's crucial to consider what type will achieve the desired effect, while naturally suiting the existing landscape. It's also critical to ensure the landscape can support a water feature and that appropriate construction methods are used to protect the feature from the environment and vise versa.

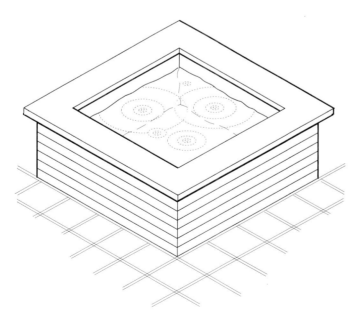

Ⓐ **RAISED POOL**
(CONCRETE, BRICK, STONE, OR WOOD)

WATER-FEATURE TYPES—While structurally similar, water features are categorized for clarity's sake. Depending on the defining characteristics of each type, a feature may or may not contain every structural component outlined in this chapter, but the "guts" of each are the same. For clear communication between builder, supplier, and client, here are some standard definitions of water-feature types.

Ponds—A pond is a pool of water that's typically contained on all sides. Its water can be moving or static. Movement can be generated by a pump, a spray jet or fountain within the pond, or water moving from a connected element, such as a waterfall. Bogs, often found at the edge of ponds, are structurally similar (and potentially connected) to the pond. Bogs are zones of highly saturated soil that support certain types of plants (see 168A).

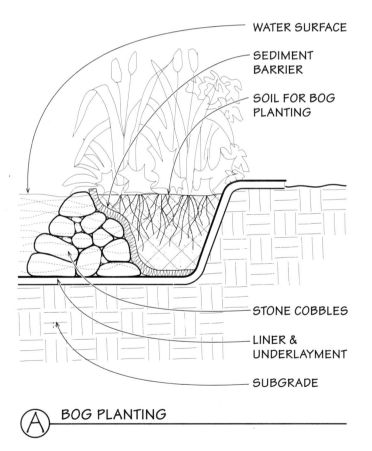

WATER SURFACE

SEDIMENT BARRIER

SOIL FOR BOG PLANTING

STONE COBBLES

LINER & UNDERLAYMENT

SUBGRADE

Ⓐ BOG PLANTING

Waterfalls—Waterfalls are defined as elements that transfer water from a higher elevation to a lower one. Water is either dropped directly from one height to another or it can flow over a series of small drops, such as those created by an artistic arrangement of boulders.

Fountains—Fountains, which span a wide array of water features, usually are more formal, even geometric, in their design, and almost always incorporate moving water of some sort. For the purposes of this book, fountains will be defined as formal water features that incorporate moving water in the form of jets, sprays, or water flowing over walls, steps, and sculpted weirs.

Streams—Streams are gradually sloping depressions that direct water through a course. They typically commence at a source pool and terminate at a splash pool downstream. Streams may be naturalistic in form, containing elements such as boulders and plants, or they may be abstracted forms that are more sculptural, made of materials ranging from concrete to tile and brick.

Water Movement—If and how a feature will incorporate moving water is an important design consideration. Water movement in a feature can take on many different forms, from a skyward spray to a downward flow. There are myriad spray nozzles, including geyser fountains, bubblers, and tulip sprays, and each lends a unique characteristic to the water's movement, creating different looks and sounds (see 169A).

Moving water has great visual and auditory impact. The nature of its look and sound is affected by many things, including the depth of the water flowing over the edge (the more water, the deeper the sound, and the more disrupted the water will appear); the height of the fall (the longer the drop, the louder the sound, and the more visual disruption); the nature of the surface that the water is flowing over (rough produces a foamy appearance while smooth gives a glasslike appearance); and the depth of the pool into which a fall empties (the deeper the pool, the deeper the sound) (see 169A to 170D).

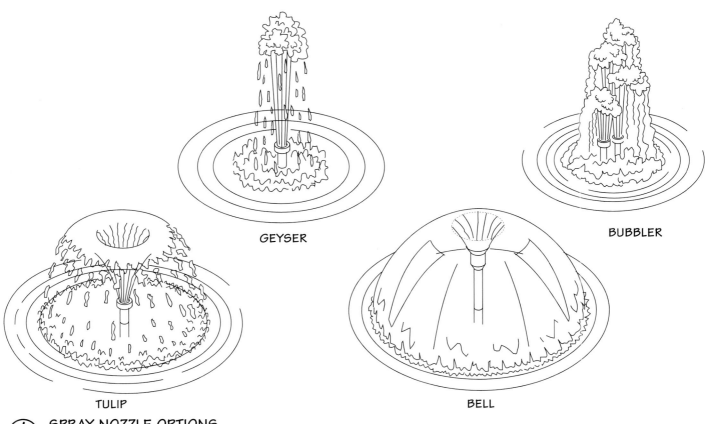

GEYSER

BUBBLER

TULIP

BELL

Ⓐ SPRAY NOZZLE OPTIONS

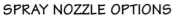

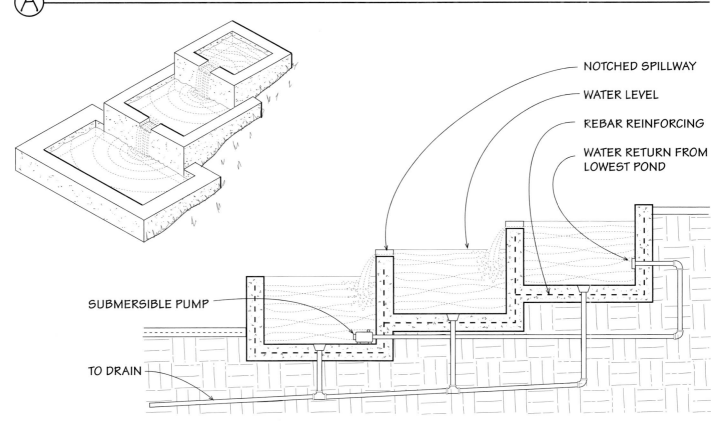

NOTCHED SPILLWAY

WATER LEVEL

REBAR REINFORCING

WATER RETURN FROM LOWEST POND

SUBMERSIBLE PUMP

TO DRAIN

Ⓑ TIERED FOUNTAIN SCHEMATIC

SHARP CORNER ALLOWS SOME WATER IN A SHALLOW FLOW TO FALL OUT FROM SURFACE, WHILE STILL MAINTAINING SOME FLOW DOWN FACE OF WALL.

STEEL OR PLEXIGLASS EDGE ATTACHED TO CONCRETE

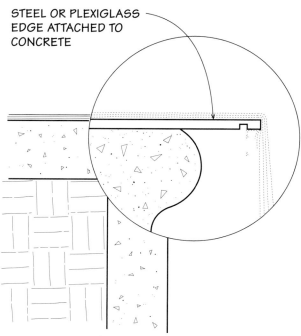

EXTENDED EDGE MOVES WATER AWAY FROM SURFACE, ENCOURAGING A SHEET FLOW OVER EDGE WITHOUT DEEP POOL ABOVE FALL. NOTE THE NOTCH ON UNDERSIDE OF OVERHANG, WHICH PREVENTS WATER FROM TRAVELING BACK UNDER MATERIAL.

Ⓐ SQUARE NOSE

Ⓑ EXTENDED NOSE

WATER

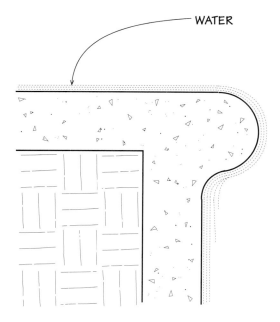

ROUNDED EDGE ALLOWS SHALLOW FLOWS TO ADHERE TO SURFACE.

POINTED WEIR CREATES POOL ABOVE FALL & ENCOURAGES SHEET FLOW TO EXTEND AWAY FROM VERTICAL SURFACE.

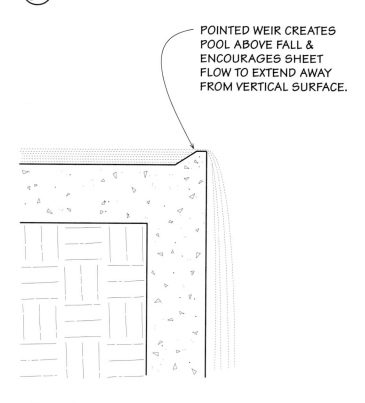

Ⓒ ROUNDED NOSE FOR FALL

Ⓓ POINTED WEIR

WATER-FEATURE PLACEMENT—Selecting the proper location for the water feature is a critical step in the design process and can impact the level of maintenance required over time, as well as the health and longevity of the feature itself. The placement of the feature should be a function of several considerations.

Codes—Local codes vary greatly, particularly in regards to how close to property lines or easements a water feature can be built. Other codes may determine how close to a structure or sensitive natural area a feature can be placed.

Sunlight—Placing a water feature in direct sunlight can encourage a higher level of algae growth and make the water temperature unsuitable for many fish and plant species. Selecting a shaded or partially shaded location will help mitigate these problems.

Vegetation—Having large, established plants can be a real asset in the design of a water feature. Care should be taken, however, to respect the root zone of large trees and shrubs since digging within the drip line can harm established vegetation severely (see Chapter 1, p. 15). Depending on the type of vegetation, leaf litter could be an issue, as well as damage from falling branches.

Visibility—Water features can create beautiful views, but it's important to consider how a feature will best fit into the landscape and what, if any, activities it will support or interfere with. Ponds or fountains in proximity to basketball courts are not ideal, for instance, while a fountain adjacent to a patio area may be a perfect addition to the garden.

Environmental conditions—Consulting local nurseries and contractors about the challenges and benefits of water features in specific regions can assist in the design and decision-making process. In areas where drought conditions are regular occurrences, for example, many water features have become planters over time due to water-conservation needs. Local animals also need to be considered, as many will make liberal use of your water feature, given the opportunity. Deer, for instance, love to wade in shallow water, and their hooves can do considerable damage to the feature's shell.

MATERIALS AND COMPONENTS—Regardless of the style and type of water feature installed, the basic structure is the same. There are two critical challenges that a water feature's design must meet: generating a source for the water (typically done by recirculating the water) and keeping the water contained.

To maintain their form, water features utilize structural shells to seal in the water. This shell, of course, must be waterproof and must be structurally sound enough to prevent the weight of the water from deforming the feature.

In-ground features have the benefit of using the strength of the soil itself to resist at least some of the pressure, but aboveground features must rely on structurally reinforced walls to both create and maintain the feature's form (see 172A & 174C).

Rigid Liners—Historically, most landscape water features, particularly ponds, were made from reinforced concrete. Today they come in many different forms and are made from several different materials, both rigid and flexible. Rigid liners are those that maintain a consistent shape and are not as subject to deformation under weight (175A).

Concrete and masonry—Concrete is still used today. It provides many options for surface finishes and can be used to create dramatic and customized above-ground or in-ground water features. It's particularly suited to formal pool or fountainlike forms. However, since concrete is highly alkaline, it is toxic to fish and many plants; a neutralizer is necessary if the feature will contain fish.

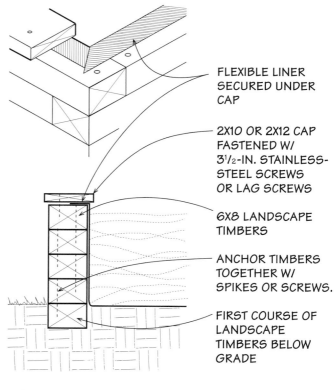

FLEXIBLE LINER SECURED UNDER CAP

2X10 OR 2X12 CAP FASTENED W/ 3½-IN. STAINLESS-STEEL SCREWS OR LAG SCREWS

6X8 LANDSCAPE TIMBERS

ANCHOR TIMBERS TOGETHER W/ SPIKES OR SCREWS.

FIRST COURSE OF LANDSCAPE TIMBERS BELOW GRADE

Ⓐ **RAISED POOL**
TIMBER WALL ENCLOSURE

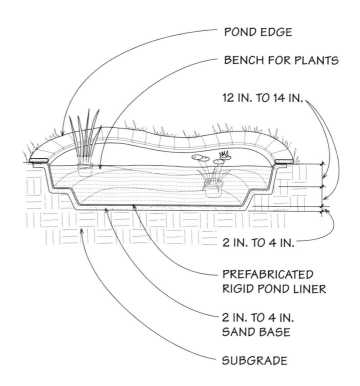

POND EDGE

BENCH FOR PLANTS

12 IN. TO 14 IN.

2 IN. TO 4 IN.

PREFABRICATED RIGID POND LINER

2 IN. TO 4 IN. SAND BASE

SUBGRADE

Ⓑ **TYP. PREFABRICATED RIGID POND INSTALLATION**

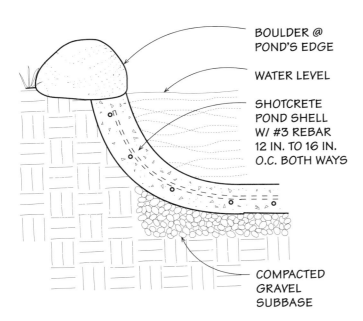

BOULDER @
POND'S EDGE

WATER LEVEL

SHOTCRETE
POND SHELL
W/ #3 REBAR
12 IN. TO 16 IN.
O.C. BOTH WAYS

COMPACTED
GRAVEL
SUBBASE

NOTE: SEAL CONCRETE W/ APPROVED
AQUATIC SEALER.

 SHOTCRETE POND

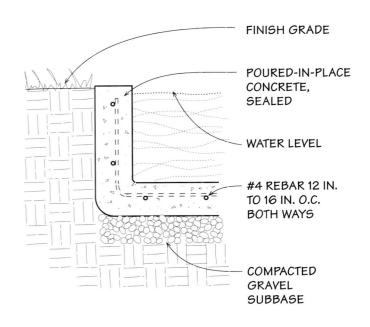

FINISH GRADE

POURED-IN-PLACE
CONCRETE,
SEALED

WATER LEVEL

#4 REBAR 12 IN.
TO 16 IN. O.C.
BOTH WAYS

COMPACTED
GRAVEL
SUBBASE

NOTE: SEAL CONCRETE W/ APPROVED
AQUATIC SEALER.

 POURED-IN-PLACE FEATURE

Concrete can be used either as a poured-in-place form or as shotcrete, which is a thick concrete material that is shot or blown into place, then trowel finished (see 173A & B). It must incorporate some form of reinforcement, such as rebar, to help control cracking. Various products, ranging from vinyl concrete patch to special paints and sealers, also will help control cracking.

Other types of masonry also may be used to form the shell of the feature, including mortared brick and concrete masonry units (CMUs). However, CMUs must be lined or sealed because they do not hold water well over time on their own (see 173C, 174A, B, & C).

Preformed polyethylene shells—Preformed polyethylene shells are another form of rigid liner. Pressed from a mold, these ¼-in.-thick, preformed shells come in a variety of shapes and sizes. They often have built-in plant shelves and a central well for a pump. They are most commonly black, but a range of colors is becoming more available.

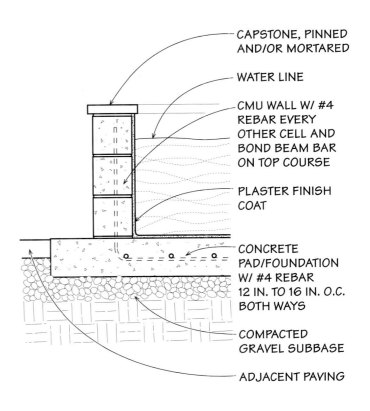

CAPSTONE, PINNED
AND/OR MORTARED

WATER LINE

CMU WALL W/ #4
REBAR EVERY
OTHER CELL AND
BOND BEAM BAR
ON TOP COURSE

PLASTER FINISH
COAT

CONCRETE
PAD/FOUNDATION
W/ #4 REBAR
12 IN. TO 16 IN. O.C.
BOTH WAYS

COMPACTED
GRAVEL SUBBASE

ADJACENT PAVING

 CMU POND W/ PLASTER LINER

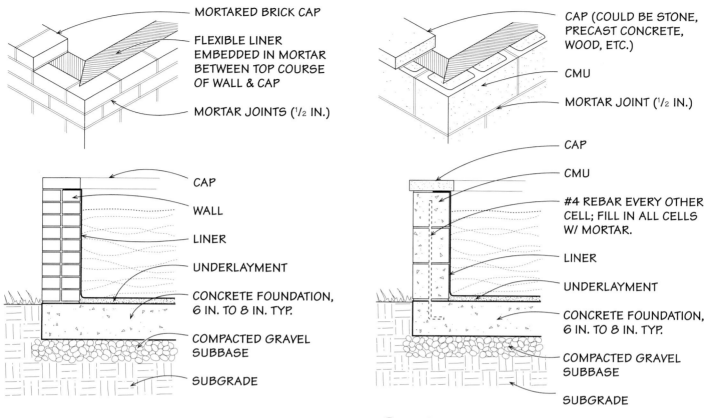

MORTARED BRICK CAP

FLEXIBLE LINER EMBEDDED IN MORTAR BETWEEN TOP COURSE OF WALL & CAP

MORTAR JOINTS (½ IN.)

CAP

WALL

LINER

UNDERLAYMENT

CONCRETE FOUNDATION, 6 IN. TO 8 IN. TYP.

COMPACTED GRAVEL SUBBASE

SUBGRADE

Ⓐ RAISED POND W/ FLEXIBLE LINER
BRICK-WALL ENCLOSURE

CAP (COULD BE STONE, PRECAST CONCRETE, WOOD, ETC.)

CMU

MORTAR JOINT (½ IN.)

CAP

CMU

#4 REBAR EVERY OTHER CELL; FILL IN ALL CELLS W/ MORTAR.

LINER

UNDERLAYMENT

CONCRETE FOUNDATION, 6 IN. TO 8 IN. TYP.

COMPACTED GRAVEL SUBBASE

SUBGRADE

Ⓑ RAISED POND W/ FLEXIBLE LINER
CMU ENCLOSURE

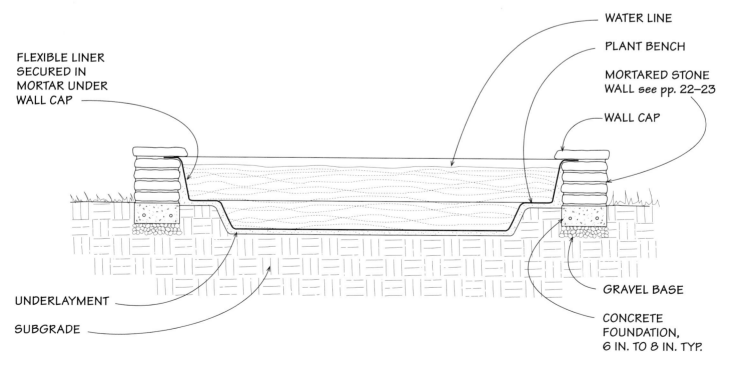

FLEXIBLE LINER SECURED IN MORTAR UNDER WALL CAP

WATER LINE

PLANT BENCH

MORTARED STONE WALL see pp. 22–23

WALL CAP

UNDERLAYMENT

SUBGRADE

GRAVEL BASE

CONCRETE FOUNDATION, 6 IN. TO 8 IN. TYP.

Ⓒ RAISED POOL W/ FLEXIBLE LINER
STONE-WALL ENCLOSURE

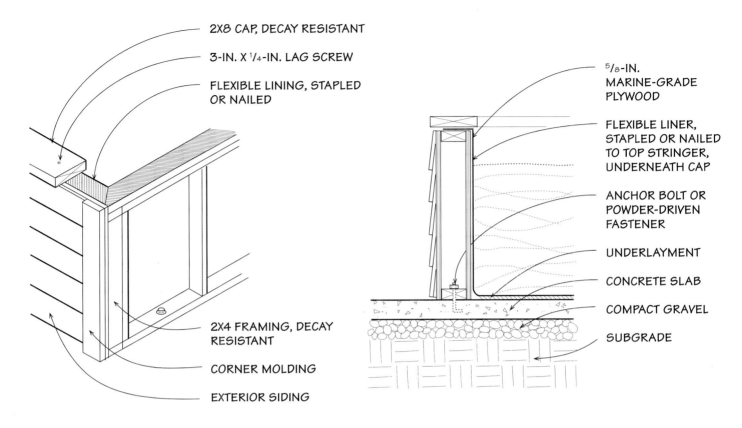

2X8 CAP, DECAY RESISTANT

3-IN. X ¼-IN. LAG SCREW

FLEXIBLE LINING, STAPLED OR NAILED

$^5/_8$-IN. MARINE-GRADE PLYWOOD

FLEXIBLE LINER, STAPLED OR NAILED TO TOP STRINGER, UNDERNEATH CAP

ANCHOR BOLT OR POWDER-DRIVEN FASTENER

UNDERLAYMENT

CONCRETE SLAB

COMPACT GRAVEL

SUBGRADE

2X4 FRAMING, DECAY RESISTANT

CORNER MOLDING

EXTERIOR SIDING

 RAISED POND

WOODEN ENCLOSURE ON CONCRETE SLAB

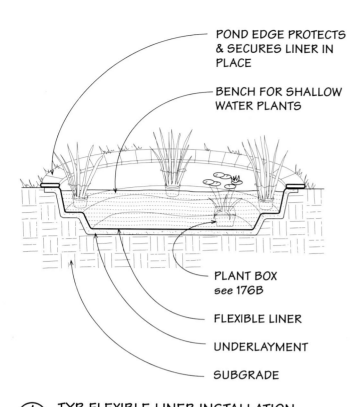

POND EDGE PROTECTS & SECURES LINER IN PLACE

BENCH FOR SHALLOW WATER PLANTS

PLANT BOX
see 176B

FLEXIBLE LINER

UNDERLAYMENT

SUBGRADE

Ⓐ   TYP. FLEXIBLE LINER INSTALLATION

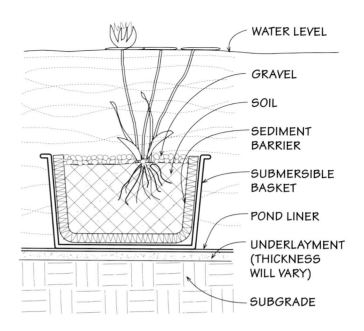

WATER LEVEL

GRAVEL

SOIL

SEDIMENT BARRIER

SUBMERSIBLE BASKET

POND LINER

UNDERLAYMENT (THICKNESS WILL VARY)

SUBGRADE

NOTE: ALSO COULD BE CERAMIC POT, BUT BASE SHOULD BE WIDE ENOUGH TO ALLEVIATE DANGER OF OVERTURN.

Ⓑ   UNDERWATER PLANT BASKET

Flexible liner materials—Shells can be made of flexible material as well. Flexible liners are those that come in large, pliable sheets of rubberized, UV-resistant material that can take the shape of nearly any excavated or built form (see 176A & C).

*Polyethylene—*A lightweight, inexpensive, and easy to work with material, polyethylene, unfortunately, is short-lived. Because it is so lightweight (6 mil. thick), polyethylene is easily punctured, and it is susceptible to rapid breakdown due to the sun's ultraviolet light.

*PVC and PVC-E—*A middle-cost option, these 20 mil. PVC (polyvinyl chloride) liners typically last 10 years or longer. They are relatively easy to work with and are less prone to puncture than both the polyethylene and rubber liners. PVC-E (where the "E" stands for "enhanced") is more supple than regular PVC, making it even easier to work with.

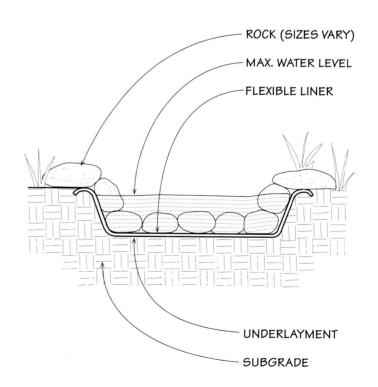

ROCK (SIZES VARY)

MAX. WATER LEVEL

FLEXIBLE LINER

UNDERLAYMENT

SUBGRADE

Ⓒ   STREAM W/ FLEXIBLE LINER

EPDM—The longest-lasting and most expensive option, this material is typically sold in 45 mil. thicknesses. Due to their thickness, however, they're also a bit more difficult to work with, particularly in irregular configurations. Care should be taken to ensure that the EPDM (which stands for "ethylene propylene diene monomer") liner is safe for use in water features, particularly those that will house fish. Some EPDM materials, such as those used in swimming pools or roofing applications, may contain chemicals that are harmful to aquatic life.

Clay—If making a very naturalistic pond is the goal, clay is a wonderful material, and it is perhaps the best option for very large ponds or for those that will be fed by a natural stream or spring. The use of clay should be considered carefully, however, due to the amount of work involved in both construction and maintenance. As the pond matures, leakage can be an issue, but the biggest potential drawback is the likelihood of tree roots or digging animals puncturing the liner.

The clay must be at least 6 in. thick and should be very pure (very little organic matter or rock and sand in the mix). A special type of clay called bentonite can be added to the surface of the clay soil to help seal the pond. Bentonite swells significantly when it comes in contact with water, providing an excellent leakage barrier. Its use, however, does necessitate that the pool remain full at all times to prevent shrinking and cracking.

Underlayment—Regardless of whether a rigid or flexible liner is used, underlayment is a critical element in the design and construction of water features because it helps ensure the longevity of the membrane enclosure. With preformed polyethylene, several inches of sand are carefully packed beneath the shell to provide support and defense against cracking. Flexible liners, such as PVC or EPDM, typically utilize a layer of old carpet or other tough material as underlayment to prevent punctures from rocks and roots. Clay liners and concrete ponds do not need any sort of underlayment.

Edge—The edge of the water feature is essentially the point at which the landscape meets the water. In a raised feature, the edge consists of the walls of the enclosure itself; in an in-ground feature, the edge must perform the role of keeping the liner (rigid or flexible) protected where it meets the surface (see 177A).

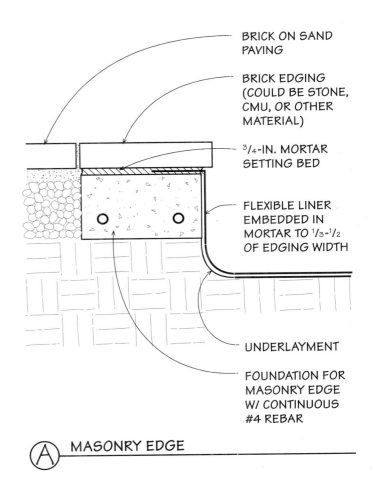

BRICK ON SAND PAVING

BRICK EDGING (COULD BE STONE, CMU, OR OTHER MATERIAL)

3/4-IN. MORTAR SETTING BED

FLEXIBLE LINER EMBEDDED IN MORTAR TO 1/3-1/2 OF EDGING WIDTH

UNDERLAYMENT

FOUNDATION FOR MASONRY EDGE W/ CONTINUOUS #4 REBAR

Ⓐ MASONRY EDGE

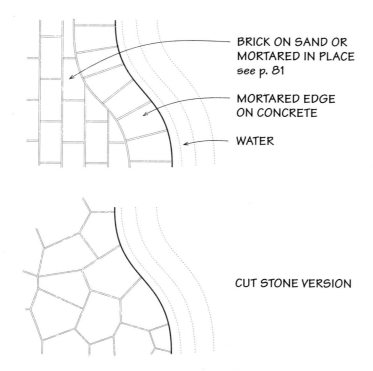

BRICK ON SAND OR
MORTARED IN PLACE
see p. 81

MORTARED EDGE
ON CONCRETE

WATER

CUT STONE VERSION

NOTE: IF USING FLEXIBLE LINER, SECURE IN
MORTAR SETTING BED BENEATH EDGE PAVING.
see pp. 81-83

 A PAVED EDGES @ POND

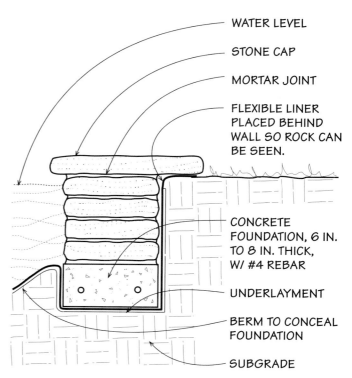

WATER LEVEL

STONE CAP

MORTAR JOINT

FLEXIBLE LINER
PLACED BEHIND
WALL SO ROCK CAN
BE SEEN.

CONCRETE
FOUNDATION, 6 IN.
TO 8 IN. THICK,
W/ #4 REBAR

UNDERLAYMENT

BERM TO CONCEAL
FOUNDATION

SUBGRADE

NOTE: ADD LAYER OF UNDERLAYMENT
ON TOP OF LINER UNDERNEATH
FOUNDATION & BEHIND ROCK WALL.

B IN-GROUND POOL W/ SUBMERGED
ROCK WALL SHELL

A more formal feature's edge may be concrete or concrete pavers (see 178A), while a more naturalistic approach might employ rocks interspersed with plants (see 178B).

Regardless of the materials chosen, a well-designed edge is critical to the aesthetics, longevity, and even safety of a water feature.

Pumps—There are two types of pumps typically used in water features—submersible and nonsubmersible. Submersible pumps range in size from 20 gallons per hour (gph) to over 4,000 gph. They are placed beneath the surface of the water, usually at the feature's lowest point (see 179A).

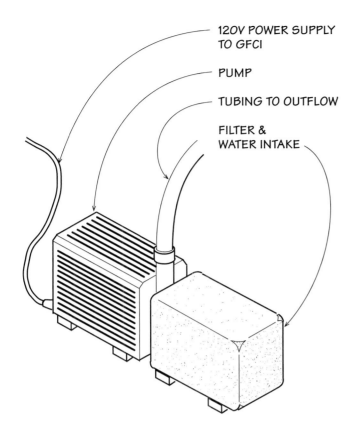

120V POWER SUPPLY
TO GFCI

PUMP

TUBING TO OUTFLOW

FILTER &
WATER INTAKE

Ⓐ SIMPLE SUBMERSIBLE PUMP W/ FILTER

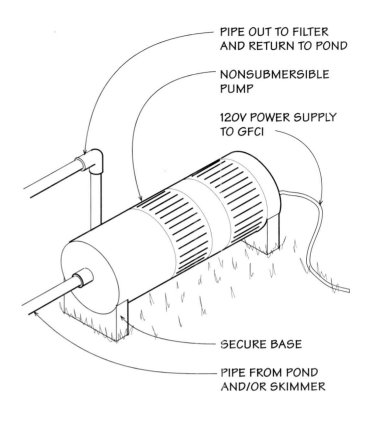

PIPE OUT TO FILTER
AND RETURN TO POND

NONSUBMERSIBLE
PUMP

120V POWER SUPPLY
TO GFCI

SECURE BASE

PIPE FROM POND
AND/OR SKIMMER

Ⓑ NONSUBMERSIBLE PUMP

Nonsubmersible pumps are placed away from the water feature, and are usually used to pump large amounts of water, often in excess of 5,000 gph (see 179B). In cold climates, pumps and other mechanical equipment such as filters and timers may be placed in below-ground vaults or boxes to help protect them from the elements. Check with a local water-feature specialist to determine the preferred or required installation.

Selecting the right pump is critical for achieving the desired effect—too little water and the sound and sight of the feature will not be sufficiently dramatic; too large and there will be too much noise and splash. The pump also needs to project the water far enough vertically (called "head," this is a measure of feet above the pump) to ensure that the feature functions properly (see the diagram at right; see tables on p. 180).

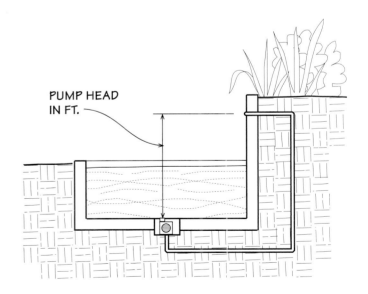

PUMP HEAD
IN FT.

Typ. Flow Rates for Water Features

Garden hose w/ no nozzle (avg. pressure)
800 gph (gallons per hour)

Ponds (pumps below are for filtration only; waterfalls, fountains, etc. in the design may
require a higher gph)

Small	40 gph to 500 gph
Medium	100 gph to 1,000 gph
Large	500 gph to 4,000 gph

Fountains
Varies due to type & size. Many commercially
available fountains run between 150 gph &
400 gph.

Splashing statuary

Small	40 gph to 150 gph
Medium	100 gph to 400 gph
Large	250 gph to 800 gph

Table adapted from *Building Waterfalls, Pools & Streams*, by
Thomas & Koogle. Meredith Books, 2002.

Typ. Flow Rates for Rectangular Weirs (GPM)

Depth of water over weir	¼"	½"	¾"	1"
Gallons per minute per linear foot of weir (GPM)	4.5	13	23	36

Typ. Flow Rate for Solid Stream Jets (GPM)

Height of spray (ft. above surface)	2	4	6
Head (ft. above pump)	3	5	8
¼" orifice	2	2.8	3.4
⅜" orifice	4	6	7
½" orifice	7	11	12

Note: Since the flow rate of many pumps is rated
in terms of Gallons Per Hour (GPH), multiply
above GPM numbers by 60 to determine approx.
pump requirements.

Table adapted from *Timesaver Standards for Landscape
Architecture*, by Harris, Charles and Dines, Nicholas.
McGraw-Hill, 1998, 2nd ed.

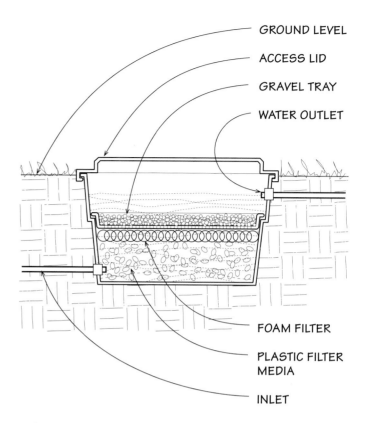

GROUND LEVEL
ACCESS LID
GRAVEL TRAY
WATER OUTLET
FOAM FILTER
PLASTIC FILTER MEDIA
INLET

 TYP. MECHANICAL FILTER

Filters—Filters are either mechanical or biological.
Most features use a mechanical filter, which collects
algae, bacteria, and other elements as the water flows
through a series of porous pads (see 180A). Biological
filtration can be as simple as having plants in the
water, or it can involve a more complex system contained in either a submerged or out-of-water tank. For
a biological filter to work properly, approximately 25
percent of the total volume of water in the feature
should circulate through the filter every hour (i.e., a
fountain containing 400 gallons of water would
require a pump that pushes 100 gph through the filter in order to keep the water clean and healthy). For
a mechanical filter, the percentage of water needing to
flow through it increases to 50 percent of the total
volume per hour (see 181A).

Skimmer—In areas where a feature is susceptible to
large quantities of leaf litter and other organic debris,
a skimmer may be a good addition to the pump and
filtration system. A skimmer removes floating debris
from a feature before it can sink and decompose,
potentially harming the water quality. This is especially

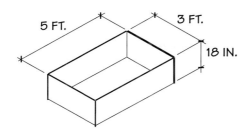

L X W X D X 7.5
5 X 3 X 1.5 X 7.5 = 169 GAL

CIRCULAR POOL

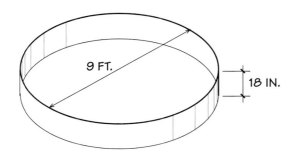

L X W X D X 6.7
15 X 6 X 2 X 6.7 = 1,005 GAL

OVAL POOL

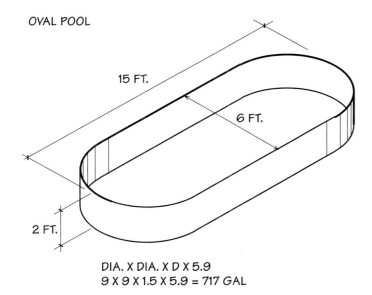

DIA. X DIA. X D X 5.9
9 X 9 X 1.5 X 5.9 = 717 GAL

NOTE: DIMENSIONS ARE EXAMPLES ONLY; SUBSTITUTE
YOUR OWN DIMENSIONS WHERE APPROPRIATE.

 CALCULATING GALLONAGE

FOR POOLS/PONDS

important in climates where ice forms on the water's surface, trapping harmful gases that can harm fish and other aquatic life.

Drain—Many features will require maintenance that involves draining the water from time to time. Unlike storm-water drains, water-feature drains must have a valve of some sort (typically a gate or ball valve) that allows the water to be released when desired. These valves are placed outside of the water feature, usually in an underground valve box that can be accessed easily. The drain ultimately connects to the storm-water system unless local codes require otherwise.

Overflow—It is always important to consider what will happen if the water feature overflows for some reason. In small features this may not be as big a concern, but for larger features, particularly those in paved areas or in proximity to a structure or existing stream, taking care of the overflow is a critical concern. In some situations, simply having a low point on the feature's edge that empties into a swale may be fine; in other circumstances, a more formal, piped system that empties into a storm-drain system may be necessary (see 182A & B). Check local codes to see what options are allowable.

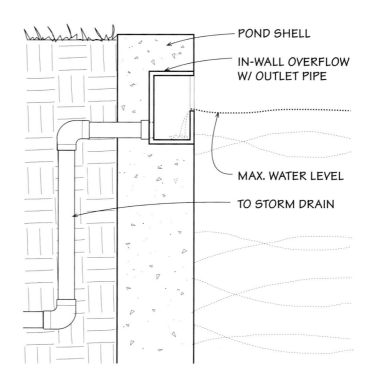

- POND SHELL
- IN-WALL OVERFLOW W/ OUTLET PIPE
- MAX. WATER LEVEL
- TO STORM DRAIN

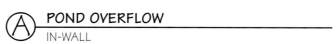

(A) **POND OVERFLOW**
IN-WALL

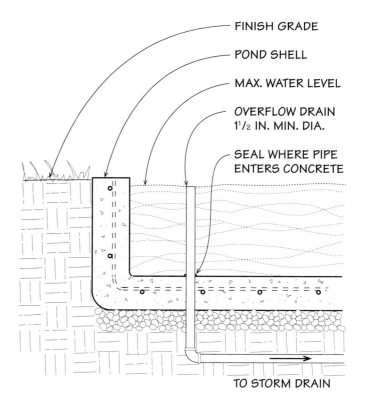

- FINISH GRADE
- POND SHELL
- MAX. WATER LEVEL
- OVERFLOW DRAIN 1 1/2 IN. MIN. DIA.
- SEAL WHERE PIPE ENTERS CONCRETE

TO STORM DRAIN

(B) **POND OVERFLOW**
STANDIPE

Water fill—Evaporation, splashing, or leakage occurs with every water feature, creating a need to refill the water from time to time. Options range from dragging out the garden hose to more formal systems involving a pipe that is connected to the domestic water source and controlled by a valve. The valve either can be a manual valve, such as a gate or ball valve (see 183A) or it can be a more complex automatic fill valve (sometimes referred to as a "float" valve), similar in some respects to the tank refill valve in a toilet. In regions with adequate, year-round rainfall, an all-natural approach with rainwater is a solution worth exploring.

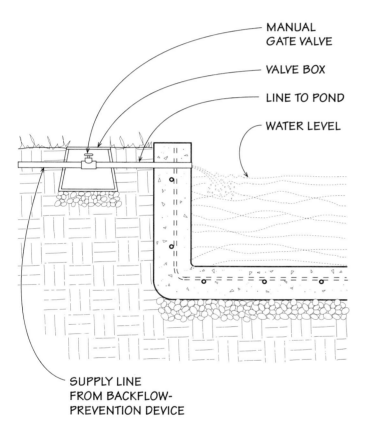

MANUAL GATE VALVE

VALVE BOX

LINE TO POND

WATER LEVEL

SUPPLY LINE FROM BACKFLOW-PREVENTION DEVICE

Ⓐ MANUAL FILL VALVE

Piping—Many different materials are available for use in water-feature plumbing. Historically, most plumbing was done with galvanized steel pipe that required a great deal of expertise to cut, thread, and install. Today, almost all plumbing for outdoor residential water features uses plastic PVC or vinyl piping.

Rigid PVC—Rigid PVC pipe is typically used for underground plumbing, since it's generally not UV-resistant, and its white color makes it difficult to hide in the landscape. Thicker PVC pipe, a class defined as schedule 40, should be used where possible, and pressurized water lines (such as a pipe run from the pump to the jet or outfall) should be installed below the frost line.

Corrugated black vinyl—Sometimes called flexible PVC, corrugated black vinyl pipe is another option that is viable for water features. While smooth like rigid PVC on the inside, its corrugated exterior allows it to bend and flex without kinking, making it ideal for use in features where piping may need to snake around roots or rocks. Unlike rigid PVC, which must be connected with gluelike solvents, corrugated black vinyl pipe is connected with stainless-steel hose clamps.

Black-vinyl tubing—Black-vinyl tube is a fairly strong material that is also flexible enough to bend around rocks and other elements. Its dark color makes it relatively easy to hide, so it works particularly well in naturalistic features that have a dark shell.

Clear-vinyl tubing—The least expensive option for this type of plumbing, clear-vinyl tubing is flexible and relatively easy to conceal as well. However, it is not as durable as black vinyl tubing, and its clear walls allow sunlight to permeate inside the pipe, encouraging algae growth. Generally, it's best used in short runs connecting underwater pond elements in shady areas.

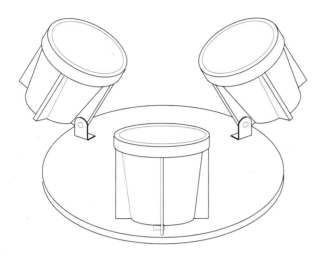

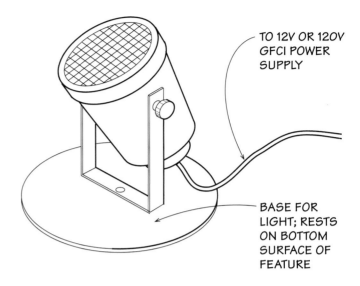

TO 12V OR 120V GFCI POWER SUPPLY

BASE FOR LIGHT; RESTS ON BOTTOM SURFACE OF FEATURE

 SUBMERSIBLE FOUNTAIN LIGHT

 SUBMERSIBLE FLOODLIGHT

Electrical hookup—Lighting and pumps require a power source. For almost all systems, a normal 120-volt line-voltage source with GFCI protection will be sufficient. Check local codes for wiring requirements before beginning construction.

Lighting—Incorporating lighting into a water feature allows it to be a dramatic element in the nighttime landscape. Both line and low-voltage options are available, as well as built-in and drop-in lighting (see 184B). Much like the landscape lighting discussed in chapter 4, water-feature lighting should accentuate interesting elements, such as statuary, special plants, or the form of the feature itself. Care should be taken to mask the source of light where possible to ensure a focus on the element being lit rather than the light itself.

COLOR CONSIDERATIONS—The color of a water feature's shell should be a major consideration in its design, as different visual effects are made possible through the selection of a light or dark surface.

Dark surfaces—Dark materials allow the surface of the water to reflect the sky and a water feature's sur-

roundings while minimizing the view into the water itself. This allows things like planters and submersible pumps to remain hidden. Both the flexible lining materials and the preformed rigid liners typically come in black; other colors, including grays, tans, and browns, are available in some of the materials.

The potential drawback of a dark color is that it can work as a heat sink, attracting and holding heat and raising the temperature of the water. This should be taken into account when the feature is designed, especially if fish are going to be incorporated.

Light surfaces—A light shell allows a view through the water to the feature's bottom. This is most typically found in swimming pools and in fountains where a tile mosaic or other adornment has been included.

There are additional possibilities that fall along the dark and light spectrum, especially if concrete is the material being used. A wide range of coloration options exist for concrete (see p.97), providing designers with a multitude of possibilities for a feature's finish.

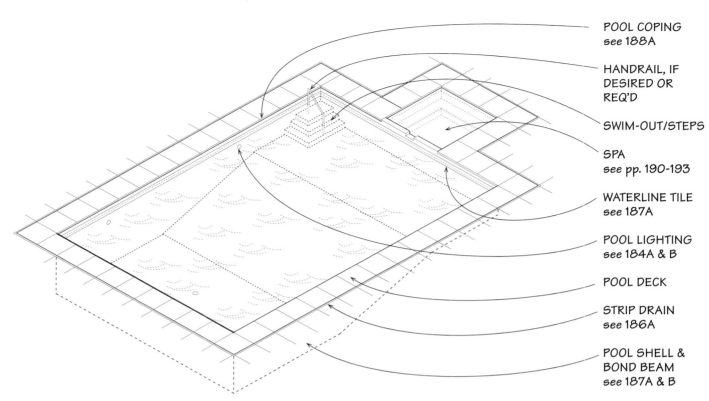

POOL COPING
see 188A

HANDRAIL, IF
DESIRED OR
REQ'D

SWIM-OUT/STEPS

SPA
see pp. 190-193

WATERLINE TILE
see 187A

POOL LIGHTING
see 184A & B

POOL DECK

STRIP DRAIN
see 186A

POOL SHELL &
BOND BEAM
see 187A & B

NOTE: FOR PLUMBING DIAGRAM, see p. 189.

POOLS AND SPAS

While water is compelling as a visual, ecological, or auditory element in the landscape, the ability to interact with it can be an even more sublime experience.

Although the options for pools and spas are numerous, the existing landscape and code or zoning requirements will frequently have a great impact on design. Due to both their scale and potential safety issues, pools and spas must be worked into the landscape with care, and in a manner that allows other desired activities to be supported as well. It's wise to consult with a waterscape specialist to ensure that a design is functional and safe, as well as aesthetically pleasing in the landscape.

IN-GROUND POOLS AND SPAS—In-ground pools
and spas share many of the same characteristics and structural needs, as well as requiring similar design considerations. When a client wants both, the water features are frequently designed as an integrated unit, sharing much of the plumbing, pumps, and filtration equipment.

In-ground pools and spas also may be discrete but adjacent elements, with no water moving from one to the other, or they may be sited completely apart.

Design considerations—There are several things to consider when designing a pool or in-ground spa, but two key issues will dictate the scope and direction of the design—the landscape and the codes.

The landscape—Properly locating an in-ground pool or spa in the landscape is perhaps the most important decision to be made in terms of creating a usable and pleasing environment. Due to the amount of excavation required, particularly for a pool, and the amount of trenching necessary for the plumbing, one major consideration will be the construction's impact on any major trees in the area. The critical root zone for many tree species is between one and one-and-a-half the diameter of the drip line (see p. 15).

Existing utilities (phone, cable, electric, gas) in proximity of the construction zone will also affect the location of a pool or spa, as will any easements or local regulations dictating setbacks from property lines, fences, or structures. Solar orientation also may be a consideration since it can aid in the passive and/or active heating of the water.

Code requirements—There are many code requirements for pools and spas that impact their design and construction. Fencing is one of the primary considerations. Many local codes require that any pool, pond, or spa that is deeper than 18 in. be enclosed by a fence (4 ft. to 6 ft., depending upon jurisdiction) with a self-closing gate (i.e., a gate that closes on its own, much like a spring on a screen or storm door). Many jurisdictions also mandate the location of the fence; some, for example, require that it be placed within 4 ft. of the pool's edge.

Other requirements include alarms on doors leading directly to the pool or spa area, which would alert a parent if a child moved into that area. These types of requirements will have a major impact on the layout of the pool or spa in the landscape and upon the visual aesthetic of the landscape as a whole, and they should be examined closely prior to construction.

Structure—In-ground pools and spas are typically constructed of a special form of concrete called gunite, or shotcrete, that is reinforced with #4 rebar laid in a 12-in. grid on the floor and sides. They have a system of pumps, filters, chemical treatment fixtures, and heaters that keep the water clean, clear, and warm. Both pools and spas typically have built-in lighting and may include elements such as ladders, steps, and seats; they also may occasionally incorporate fountains or waterfalls into their design. Many times spas have an additional air blower, and they may have additional pumps to aerate the water and move it at high speed.

Pool deck—The paved surfaces around a pool and/or spa have specific requirements that other paved surfaces do not necessarily share. Traction is extremely important as is the paving material's tendency to collect and radiate heat. Many pool decks are concrete surfaces that are either textured like a sidewalk or coated with a surfacing material similar to that which is applied to the walls of the pool. Sometimes referred to as a "cool deck," this surface disperses and reflects the heat, making it both slip resistant and comfortable to walk on. It is applied as a slurry coat over the top of hardened concrete and is textured with a special trowel.

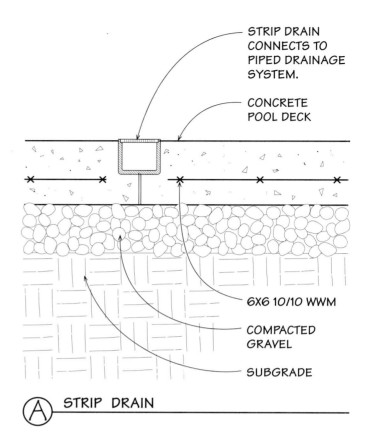

STRIP DRAIN CONNECTS TO PIPED DRAINAGE SYSTEM.

CONCRETE POOL DECK

6X6 10/10 WWM

COMPACTED GRAVEL

SUBGRADE

(A)   STRIP DRAIN

Due to its high slip factor, tile should be avoided in most circumstances, although small accent areas can be incorporated as an aesthetic element. Care should be taken to avoid placing these elements in high-traffic areas, however, because even a small area can cause a large slip. Other materials, such as brick, pavers, or flagstone paving, can be used in pool areas, but since some are as slick as tile, they should be selected carefully.

The pool deck also must be properly drained. The chemicals in pool water can adversely affect surrounding vegetation, and dirt on the pool deck can adversely affect the pool. Therefore, all drainage should be directed away from the pool's edge and should be collected by some sort of drainage structure (such as a trench drain or strip drain) (see 186A).

Wall—The pool or spa wall refers to the sides and bottom of the feature. Its thickness will range from 6 in. to 9 in., depending upon soil type and the depth of the pool or spa. It is formed with gunite and rebar, and surfaced with a plaster material that helps seal the concrete and give it a smooth finish. This surfacing material also can be colored to enhance the visual impact of the pool or spa in the landscape (see 187A).

Bond beam—The bond beam is the top portion of the pool or spa wall that forms a ring and helps it maintain its shape. Formed either with gunite and rebar or with concrete that is cast in place, the bond beam is typically 12 in. to 14 in. in width and extends approximately 8 in. to 12 in. below the waterline. Its height above the waterline can vary greatly since it has become very common to use the bond beam as a retaining wall, extending it structurally above the waterline and the elevation of the pool deck. This sort of treatment allows plants and/or grade change to be addressed in some very creative ways. The top of the bond beam is typically covered by the coping (see 187B).

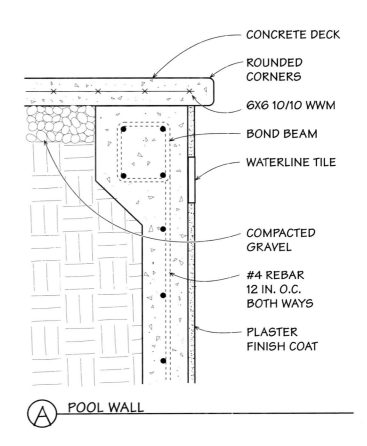

CONCRETE DECK
ROUNDED CORNERS
6X6 10/10 WWM
BOND BEAM
WATERLINE TILE
COMPACTED GRAVEL
#4 REBAR 12 IN. O.C. BOTH WAYS
PLASTER FINISH COAT

Ⓐ POOL WALL

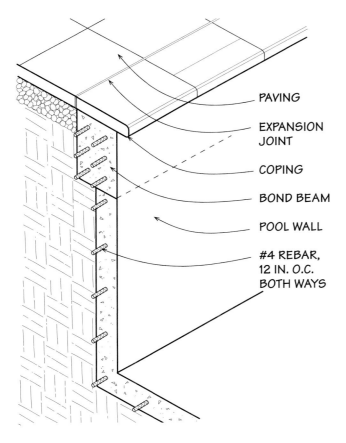

PAVING
EXPANSION JOINT
COPING
BOND BEAM
POOL WALL
#4 REBAR, 12 IN. O.C. BOTH WAYS

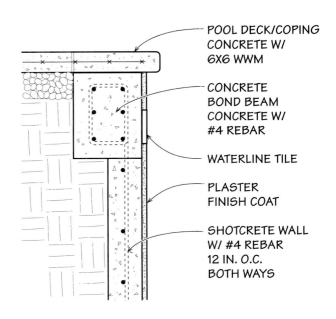

POOL DECK/COPING CONCRETE W/ 6X6 WWM
CONCRETE BOND BEAM CONCRETE W/ #4 REBAR
WATERLINE TILE
PLASTER FINISH COAT
SHOTCRETE WALL W/ #4 REBAR 12 IN. O.C. BOTH WAYS

Ⓑ BOND BEAM

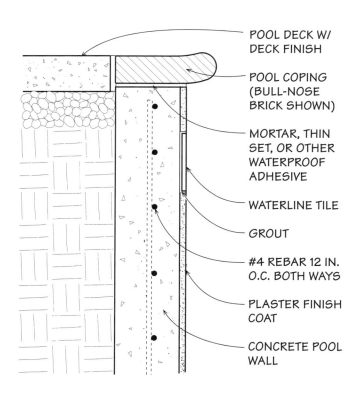

POOL DECK W/
DECK FINISH

POOL COPING
(BULL-NOSE
BRICK SHOWN)

MORTAR, THIN
SET, OR OTHER
WATERPROOF
ADHESIVE

WATERLINE TILE

GROUT

#4 REBAR 12 IN.
O.C. BOTH WAYS

PLASTER FINISH
COAT

CONCRETE POOL
WALL

 COPING—PRECAST CONCRETE OR BRICK

Coping—The coping is the material that surrounds the pool or spa at the edge of the paving, differentiating it from the rest of the decking. In some cases, the coping is a special material, such as brick, precast concrete, stone, or tile. In other cases, the decking itself extends out over the bond beam and is "edged" or "tooled" (where the corners are rounded off to make the edge less sharp), creating a seamless deck that extends to the pool or spa's edge (see 188A).

Waterline tile—This is a ring of tile that's typically located about 6 in. to 9 in. below the top of the bond beam at the waterline. Many of the oils and particles that fall into a pool/spa float for a time on the surface and then collect on the walls; this ring of tile creates a surface that is much easier to clean and care for than the plaster surface of the pool. These tiles are set on the concrete surface of the bond beam and secured with waterproof adhesive and grout (see 188A).

Swim-out/Walk-out—Every pool and spa must have at least one exiting option that is easier and safer than climbing over the edge. It is important to note that ladders alone may not meet local requirements for this purpose, although their inclusion in pool design also may be mandated by local codes.

In pools, a swim-out or walk-out often takes the form of a staircase and railing in the corner of the pool's shallow end, but it also can be designed as a ramp, allowing access for people who use wheelchairs or for those who prefer a gradual slope rather than steps. Spas often will use the built-in seats as a step down into the water, but a railing is generally required to ensure safe access and egress. Whatever form the swim-out or walk-out takes, it is imperative that it be carefully incorporated into the pool or spa's design to ensure that it does its job while not creating its own safety hazard via bad placement.

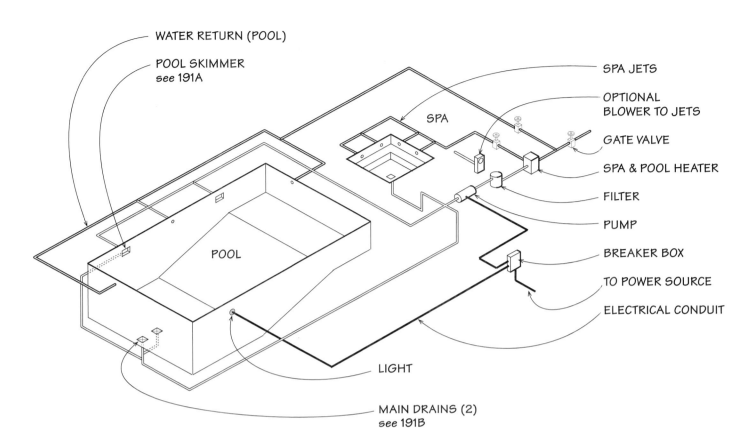

WATER RETURN (POOL)

POOL SKIMMER
see 191A

SPA JETS

OPTIONAL
BLOWER TO JETS

GATE VALVE

SPA

SPA & POOL HEATER

FILTER

PUMP

BREAKER BOX

TO POWER SOURCE

ELECTRICAL CONDUIT

POOL

LIGHT

MAIN DRAINS (2)
see 191B

Plumbing—The plumbing in a pool or spa is essentially a loop system. Water flows from a pump and filter to a heater. From the heater, the water flows through PVC piping (with solvent-welded, or glued, fittings capable of withstanding the chlorinated water and direct burial) to the pool or spa, where it is dispersed through several jets located below the water-line tile. Ideally, the jets are placed in a pattern that will circulate the water effectively, heating it more evenly.

Water is returned to the pump, filter, and heater through the skimmer and main drain. These plumbing systems can be relatively simple, incorporating only the basic elements, or they also can include devices that automatically balance the water's chemical makeup, and/or power the air jets in a spa. Pools and spas often will share the system, but spas may need additional pumps and blowers.

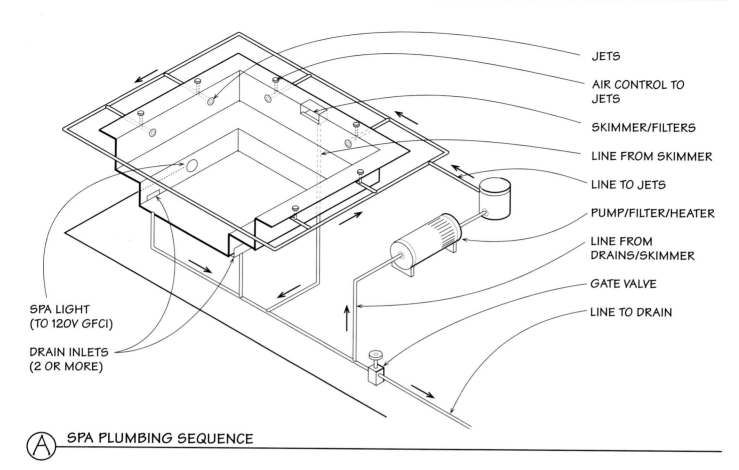

JETS

AIR CONTROL TO JETS

SKIMMER/FILTERS

LINE FROM SKIMMER

LINE TO JETS

PUMP/FILTER/HEATER

LINE FROM DRAINS/SKIMMER

GATE VALVE

LINE TO DRAIN

SPA LIGHT (TO 120V GFCI)

DRAIN INLETS (2 OR MORE)

(A) SPA PLUMBING SEQUENCE

The codes in most jurisdictions require that pumps, filters, heaters, and the system's power source be located at least 10 ft. from the pool and placed behind a fence or other barrier for safety. In most areas, the pumps will operate on a 120-volt power source, while the heater may require either a 120- or 240-volt service, depending on the type of heater being used. Local code requirements also should be considered when selecting individual elements of the plumbing system.

Pumps—Pumps should be sized so that they are capable of recirculating the entire volume of water in the pool every six to eight hours. The size of a pool will dictate the size of the pump required, and any additional features, such as a spa or waterfall, in all likelihood, will require additional pumps.

Filters—The three most common types of filters are cartridge, sand, and diatomaceous-earth filters. These mechanical filters are similar in form and function to those described earlier in the chapter. Selection

should be based on the clarity of water desired and the level of maintenance the homeowner is willing and able to support. Consult with a waterscape specialist in your area to determine the best filtration option for your project.

Heaters—The three major types of heaters are gas, electric, and solar. Gas heaters (either propane or natural gas) will heat a pool or spa the fastest. Equipment costs are generally less than electric or solar options, but operating costs may be higher in some regions, depending upon gas availability.

Electric heaters cost more to purchase and install, but their operating cost is typically lower than gas heaters, although this, too, may vary by region.

A third option is a solar heater, which uses heat-absorbing black piping or solar panels to heat the water. This option can be expensive to install and can take up much more space than conventional heaters. Once installed, however, a solar heating system is extremely inexpensive to operate.

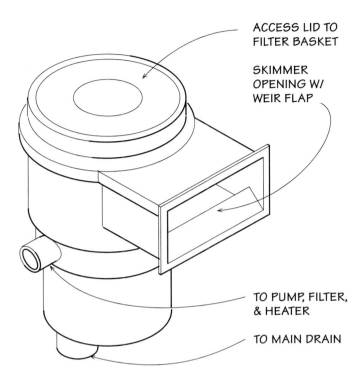

ACCESS LID TO
FILTER BASKET

SKIMMER
OPENING W/
WEIR FLAP

TO PUMP, FILTER,
& HEATER

TO MAIN DRAIN

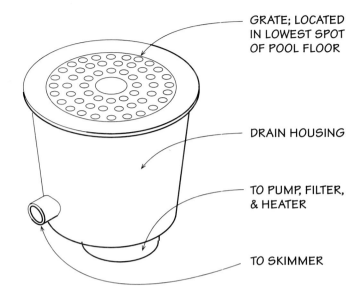

GRATE; LOCATED
IN LOWEST SPOT
OF POOL FLOOR

DRAIN HOUSING

TO PUMP, FILTER,
& HEATER

TO SKIMMER

NOTE:
INCLUDE TWO (2) DRAINS TO REDUCE
RISK OF DROWNING DUE TO SUCTION
FROM SINGLE DRAIN.

 POOL SKIMMER

 MAIN DRAIN

Because a spa is heated to a higher temperature than a swimming pool, a heater equipped to simultaneously heat a pool and spa to different temperatures will be required if a spa is built in conjunction with a pool.

Skimmer—The skimmer is typically installed at the same level as the waterline tile, and it helps remove debris from the surface of the water, providing a first line of filtration in the recirculation system (see 191A). Water flows through the skimmer, returning to the pump, where it is recirculated through the filtration and heating system.

Main drain—Water is also recirculated through the main drain, which is positioned at the lowest spot in the pool. Aside from being the means by which a pool may be drained, it serves as a deep-water return back to the pump, filter, and heater. Typically, there are two main drains installed so that the suction of water being drawn back into the system is dissipated, reducing the risk of a swimmer becoming trapped by the suction on the bottom of the pool (see 191B).

ABOVEGROUND SPAS—Aboveground spas, which include portable spas and hot tubs, are a less costly (in terms of both time and resources) alternative to the in-ground spas discussed earlier. There are some important considerations, however, when selecting and siting an aboveground spa.

Portable spas—Portable spas are prefabricated, self-contained units that include all of the plumbing, heaters, pumps, and insulation inside a constructed fiberglass shell. They are generally much more energy efficient than their in-ground counterparts and have many features built into them that would be very difficult to replicate in an in-ground spa. Lounge seating, special therapy chairs and jets, and foot massaging units are just a few of the options available. Their size will vary greatly, depending upon the number of people they are designed to hold (three to eight is the typical range).

Power source—Portable spas are essentially ready to go, needing only a power source. The power source is typically a 240-volt GFCI-protected line. An in-line breaker switch for interrupting the power to the spa (for maintenance or in case of an emergency) should be located between 5 ft. and 50 ft. away from the spa unit itself. Check local codes for requirements or necessary permits.

Foundation—The weight of the spa, once it is filled with water, is much greater than the typical load that decks and patios are designed to withstand. Great care should be taken to design and prepare a structurally sound surface for the spa. If the spa is being placed upon a concrete surface, increasing the thickness of the concrete to 6 in. and reinforcing it with welded wire mesh or rebar should be sufficient to support the weight of the spa.

If you are considering placing the spa on a wooden deck, check with local building codes for load requirements to ensure that the substructure of the deck can withstand both the weight of the water and the side-to-side rocking motion that the water could exert upon the structure.

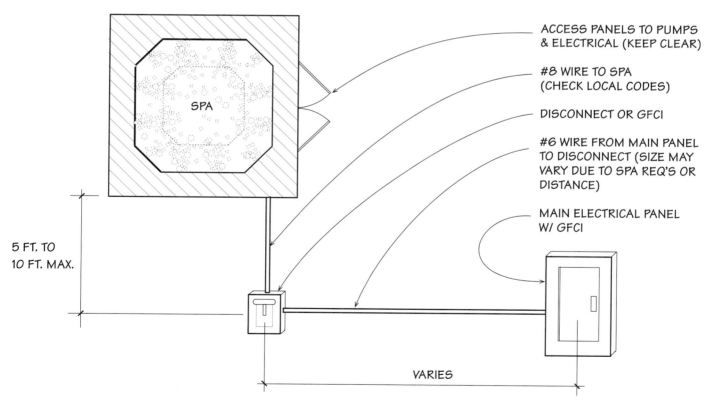

SPA

ACCESS PANELS TO PUMPS & ELECTRICAL (KEEP CLEAR)

#8 WIRE TO SPA (CHECK LOCAL CODES)

DISCONNECT OR GFCI

#6 WIRE FROM MAIN PANEL TO DISCONNECT (SIZE MAY VARY DUE TO SPA REQ'S OR DISTANCE)

MAIN ELECTRICAL PANEL W/ GFCI

5 FT. TO 10 FT. MAX.

VARIES

NOTE: CHECK LOCAL CODES FOR SPECIFIC REQ'S

 WIRING FOR PORTABLE SPAS

Placing the spa on soil is not recommended since the substructure of portable spas is often made of non-preservative-treated wood; placing it on compacted gravel may be an option, although the likelihood of ground shift over time will place the spa out of level fairly quickly. Because of this, gravel is also not generally recommended.

Placement in the landscape—Placing a portable spa in the landscape without having it look awkward or out of place can be tricky. Several options exist to mitigate this challenge, such as sinking the spa into a deck, or building the spa into a hillside where it takes on a less visible profile. Care should be taken, however, to ensure that the access panel, located on the spa's side, is always easily accessible. This side is generally removable, but some clearance is needed to take it off and to provide access to the pumps, heaters, and plumbing located inside.

Hot-tub spas—Hot-tub spas are in some ways a hybrid of portable spas and in-ground spas. The tub itself, typically constructed out of redwood or cedar, is placed above grade on a slab, deck, or gravel pad. However, the plumbing is essentially the same as it is for an in-ground spa, with jets hooked up to a pump, filter, and heater that are located away from the tub.

Piping is generally direct-burial PVC, although flexible, UV-resistant piping may be used where the piping encircles the spa. The heater and pumps are typically controlled through the use of an "air switch" located adjacent to the tub that turns the pump off and on without the need for an electrical switch or keypad.

Because hot tubs contain no additional insulation and are located above grade, they are generally the least energy-efficient type of spa. They are also more like in-ground spas in terms of limited seating options, with simple benches and a central foot well being the general rule.

LEGEND

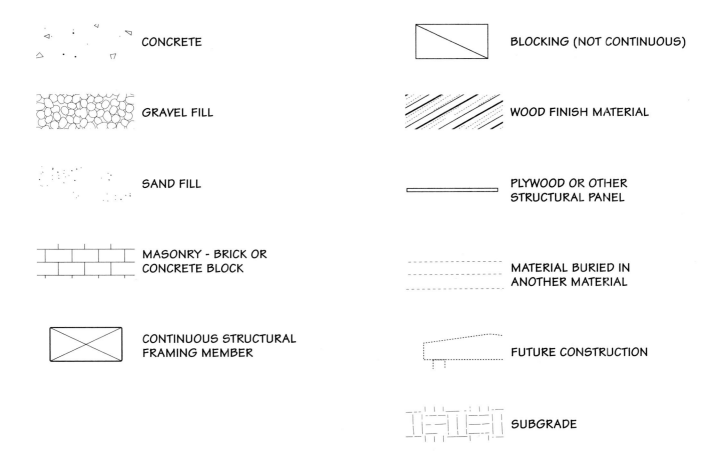

CONCRETE

GRAVEL FILL

SAND FILL

MASONRY - BRICK OR
CONCRETE BLOCK

CONTINUOUS STRUCTURAL
FRAMING MEMBER

BLOCKING (NOT CONTINUOUS)

WOOD FINISH MATERIAL

PLYWOOD OR OTHER
STRUCTURAL PANEL

MATERIAL BURIED IN
ANOTHER MATERIAL

FUTURE CONSTRUCTION

SUBGRADE

LIST OF ABBREVIATIONS

&	AND		REBAR	REINFORCING STEEL
@	AT		REQ'D	REQUIRED
APPROX.	APPROXIMATE(LY)		SQ. FT.	SQUARE FOOT/FEET
DIA.	DIAMETER		T&G	TONGUE AND GROOVE
FT.	FOOT/FEET		TYP.	TYPICAL(LY)
FFE	FINISH FLOOR ELEVATION		W	WIDTH
FFL	FINISH FLOOR LEVEL		W/	WITH
GFCI	GROUND FAULT CIRCUIT INTERRUPTOR		WP	WEATHER PROOF
H	HEIGHT		WWM	WELDED WIRE MESH
IN.	INCH			
J-BOX	JUNCTION BOX			
MAX.	MAXIMUM			
MIN.	MINIMUM			
#	NUMBER			
O.C.	ON CENTER			
LB.	POUND(S)			
PSF	POUNDS PER SQUARE FOOT			
PSI	POUNDS PER SQUARE INCH			
P.T.	PRESSURE (PRESERVATIVE) TREATED			
PVC	POLYVINYL CHLORIDE			

RESOURCES

Associated Landscape Contractors of America
12200 Sunrise Valley Dr., Suite 150
Reston, VA 22091
Ph: (703) 620-6363
Fax: (703) 620-6365
www.alca.org

Associated Professional Landscape Designers
1924 North Second St.
Harrisburg, PA 17102
Ph: (717) 238-9780
Fax: (717) 238-9985
info@apld.org
www.apld.org

Brick Institute of America
11490 Commerce Park Dr.
Reston, VA 20191-1525
Ph: (703) 620-6251
Fax: (703) 620-3928
www.brickinst.org
brick@pop.erols.com

Canadian Wood Council
1400 Blair Pl., Suite 210
Ottawa, Ontario K1J 9B8
Canada
Ph: (613) 747-5544

Construction Specifications Institute
601 Madison St.
Alexandria, VA 22314
Ph: 800-689-2900
Fax: (703) 684-8436
www.csinet.org/
csimail@csinet.org

Forest Products Society
2801 Marshall Ct.
Madison, WI 53705-2295
Ph: (608) 231-1361

Interlocking Concrete Pavement Institute
1444 "I" St. N.W., Suite 700
Washington, DC 20005-6542
Ph: (202) 712-9036
Fax: (202) 408-0285

International Erosion Control Association
P.O. Box 774904
Steamboat Springs, CO 80477-4904
Ph: (800) 455-4322
Fax: (970) 879-8563
www.ieca.org
ecinfo@ieca.org

Irrigation Association
8260 Willow Oaks Corporate Dr., Suite 120
Fairfax, VA 22031
Ph: (703) 573-3551
Fax: (703) 573-1913
www.irrigation.org
membership@irrigation.org

Landscape Contractors Association
9053 Shady Grove Ct.
Gaithersburg, MD 20877
Ph: (301) 948-0810
Fax: (301) 990-9771
www.lcamddcva.org
lca@mgmtsol.com

Landscape Maintenance Association
41 Lake Morton Dr., #26
Lakeland, FL 33801
Ph: (813) 680-4008

National Arborist Association
PO Box 1094
Amherst, NH 03031-1094
Ph: (603) 673-3311
Fax: (603) 672-2613
www.natlarb.com

National Association of Home Builders
1201 15th St. N.W.
Washington, DC 20005-2800
Ph: (800) 368-5242
Fax: (202) 822-0559
www.nahb.com
m_crrier@msn.com

National Concrete Masonry Association
2302 Horse Pen Rd.
Herndon, VA 20171-3499
Ph: (703) 713-1900
Fax: (703) 713-1910
www.ncma.org

National Forest Products Association
1111 Nineteenth St. N.W., Suite 700
Washington, DC 20036
Ph: (202) 463-2700

National Gardening Association
180 Flynn Ave.
Burlington, VT 05401
Ph: (802) 863-1308
Fax: (802) 863-5962
www.nationalgardening.com

Northeastern Lumber Manufacturer's Association
272 Tuttle Rd.
Cumberland Center, ME 04021
Ph: (207) 829-6901

Southern Forest Products Association
P.O. Box 641700
Kenner, LA 70064-1700
Ph: (504) 443-4464

Western Wood Products Association
522 S.W. Fifth Ave., Suite 500
Portland, OR 97204
Ph: (503) 224-3930

FURTHER READING

Amrhein, James E. *Residential Masonry Fireplace and Chimney Handbook, Second Ed.* Los Angeles, Calif.: Masonry Institute of America, 1995.

Editors of *Fine Homebuilding. The Best of Fine Homebuilding: Porches, Decks, and Outbuildings.* Newtown, Conn.: The Taunton Press, 1997.

Fine Gardening magazine, Newtown, Conn.: The Taunton Press, bimonthly.

Gilmer, Maureen. *Water Works.* Chicago: Contemporary Books, 2002.

Harris, Charles and Dines, Nicholas. *Timesaver Standards for Landscape Architecture.* New York: McGraw Hill, 1998.

Hendrix, Howard and Straw, Stuart. *Reliable Rain, A Practical Guide to Landscape Irrigation.* Newtown, Conn.: The Taunton Press, 1998.

Landphair, Harlow and Klatt, Fred. *Landscape Architecture Construction, Third Ed.* Upper Saddle River, NJ: Prentice Hall, 1999.

McBride, Scott. *Landscaping with Wood.* Newtown, Conn.: The Taunton Press, 1999.

Moyer, Janet. *The Landscape Lighting Book.* New York: John Wiley and Sons, 1992.

Strom, Stephen and Nathan, Kurt. *Site Engineering for Landscape Architects, Third Ed.* New York: John Wiley and Sons, 1998.

Thomas, Charles and Koogle, Richard. *Building Waterfalls, Pools and Streams.* Des Moines, IA: Meredith Books, 2002.

Vandervort, Don, ed. *Building Barbecues and Outdoor Kitchens.* Menlo Park, Calif.: Sunset Publishing Corp., 2001.

GLOSSARY

Anaerobic digestion A fermentation process in the absence of air.

Angle of coverage The spread of the spray from a sprinkler head, the greatest angle being 360°.

Anti-siphon valve A type of backflow prevention device integrated into an above-grade zone valve in an irrigation system.

Arbor A simple roofless garden structure on which vines are usually grown.

Backfill Gravel or soil that is filled into an excavation to create a new finish grade.

Backflow prevention A device designed to prevent water from an irrigation system from being pulled back into the water supply if an unexpected negative pressure occurs.

Ball valve A type of valve that uses a rotating element inside the valve to control the flow of water.

Baluster A single vertical component of a balustrade.

Balustrade A protective railing at a stair or deck made of numerous vertical elements.

Barbecue An ensemble of outdoor cooking equipment including grill, countertop, sink, and/or other devices.

Batten A small board that is used to cover the gap between two large boards.

Batter The degree to which a wall leans back into a hillside, typically measured in inches back per foot of vertical rise.

Beam A large horizontal structural member spanning between two supports.

Bearing capacity The ability of soil to support a load, measured in pounds per square foot (psf).

Belvedere A roofed garden structure, usually overlooking a view.

Bentonite A specific type of clay that creates an impermeable barrier when wet, thereby providing an impermeable pond liner.

Bio bag An erosion control device used to trap sediment and prevent it from running off-site.

Blocking Framing made of small pieces running perpendicular to joists or rafters.

Bog An area of extremely wet soil, typically located adjacent to a pond.

Bond beam The extra thick and strongly reinforced band around the top of a pool enclosure.

Brace A structural device used to laterally stabilize a structure by means of triangulation.

Bubbler A type of sprinkler head that floods an area directly around itself.

Building code A set of legal restrictions that pertain to the structural integrity of buildings and the health and safety of those who use them.

Cantilever The portion of a structural member that projects beyond its support.

Carriage The structure supporting the treads and risers of a wooden stair.

Catch basin A subsurface, boxlike drainage structure that collects water from the surface and transfers it to a subsurface pipe after allowing debris to settle out in a sump area, located at the bottom of the basin.

Check dam A small dam placed in a swale that holds water back, allowing sediments to drop out and encouraging some water to percolate back into the soil.

Check valve A valve that allows water to flow in only one direction.

Cheek wall A low wall that is adjacent to the length of a ramp or flight of stairs.

Circuit breaker A device that turns off electricity to a circuit when the capacity of the circuit is exceeded.

Cistern A holding tank for water.

Clay A type of soil that drains very poorly and is typically susceptible to severe shrink and swell.

Cleat A small strip of wood or metal used to support something on the surface to which it is attached.

Closed system (drainage) A system composed of pipes and other subsurface drainage structures.

Code See Building code.

Column A vertical structural element; a post.

Column footing The spread concrete support at the base of a column.

Compaction The tamping or compressing of soil or gravel to increase its strength.

Composting toilet A device designed to decompose human waste without the use of water.

Concrete masonry unit (CMU) A type of concrete block preformed for use in retaining walls or freestanding walls. CMUs either can be used to create dry-stack gravity walls, or they can be mortared and reinforced to create an engineered wall.

Conduit A hard plastic or metal tube designed to physically protect electrical wires from abrasion or impact.

Contour interval The vertical change in elevation between contour lines.

Contour line A line of constant elevation on the ground as measured from a specific spot, or datum point.

Control joint A line either pressed or cut into the surface of a concrete slab that provides a specific place for the concrete to crack as it expands and contracts.

Controller A programming device to regulate the timing of water distribution in an irrigation system.

Coping (pool) The edge treatment at the top of the bond beam where it meets the pool deck.

Countersink To recess below the finish surface.

Course A horizontal row or layer of bricks, blocks, or timbers. Can also refer to a layer of asphalt spread out across a surface.

Crushed rock A material composed of stones that have been ground or crushed into smaller, jagged pieces, including the dust, or fines, that is also the product of grinding.

Cut The excavation and/or removal of soil from an area.

Dado A rectangular groove cut into a board.

Datum point A specific spot, the elevation of which is the basis for all other elevations noted on a site.

Dead load The weight of a structure itself.

Deck A level outdoor surface made of wood and supported above ground level.

Deck screw A self-tapping screw made for the purpose of attaching decking to joists.

Decking Boards used for the finish surface of a deck.

Demand water heater A device that heats water instantaneously as it is used.

Direct-burial cable Electrical cable rated for burial in the earth without protection.

Distribution box A small chamber that routes effluent from a septic tank to the various lines in a leach field.

Dobe A small concrete block with wires attached that is placed beneath rebar or welded wire mesh (WWM) prior to pouring concrete.

Dosing tank An extra chamber in a septic tank that houses a pump to force effluent uphill.

Drain field See Leach field.

Drain (or drop) inlet A simple opening covered by a small slotted grate that collects water from the surface and transfers it directly into a subsurface pipe.

Drip emitter A micro irrigation head that drips water in one spot.

Drip line An imaginary line on the ground that corresponds with the outside edge of a tree's canopy.

Dry well A subgrade pit that is filled with large drain rock or cobbles for the purpose of allowing water to percolate back into the subsoil.

Eave The horizontal lower edge of a roof.

Effluent The liquid that is discharged from a septic tank.

Electrical cable A collection of wires—hot, neutral, and ground—wrapped together in a single sheath or covering.

Electrical circuit A group of electrical devices, such as receptacles and lights, that is served by a single cable from an electrical panel.

Electrical panel The enclosed metal box in a residence where the wires from the electrical utility are connected to breakers at the head of each electrical circuit.

Emitter A lateral drip line in a micro-irrigation system.

End grain The exposed wood surface at the end of a board or timber.

Engineered fill Material placed on a site and then compacted to ensure that settling or other movement is minimized.

Engineered wall A wall, typically higher than 18 in., that utilizes steel reinforcement and relies on a footing and a stem wall to provide the strength required to hold back a hillside.

Ethylene Propylene Diene Monomer (EPDM) A heavy-duty, rubberized material used as a liner in many pond and stream applications.

Expansion joint A flexible element embedded in concrete that allow the slab to expand and contract without severely cracking or heaving.

Fascia Trim board at the eave of a roof or edge of a deck.

Fibermesh Small fiberglass strands mixed in with concrete that help reduce surface cracking.

Fill The importation and/or placement of soil in a given area.

Filter fabric A below-grade textile that allows the flow of water while preventing the passing of soil particles.

Finished Floor Elevation (FFE) The elevation of the floor inside a structure.

Firebox The chamber, lined with firebricks or other refractory material, in which a fire is contained in a fireplace or barbecue.

Firebrick A brick made from special clay and other materials so that its form is not adversely affected by intense heat.

Fireclay A type of heat-resistant clay.

Flatwork Concrete that is poured level or near level in a slab-like form and finished.

Flow rate The ability of water to flow through a pipe, measured in gallons per minute (gpm).

Flue A vertical passageway designed to draw smoke from a fireplace.

Flush With finish surface on the same plane.

Footing A spread concrete base in the ground designed to support a structure; The spread portion at the base of a foundation.

French drain A simple drainage structure composed of a geotextile-encased, rock-filled trench that is placed either on or just below the surface of the landscape to help facilitate drainage in that area. Sometimes a perforated pipe will also be included in the design to increase the drain's capacity.

Frieze block Blocking between rafters at the eave of a roof.

Frost line The depth at which the ground freezes in a given locality.

Gable The vertical triangular end of a roof composed of two equally pitched slopes.

Galvanized steel Steel that is coated with zinc to protect it from rusting.

Gate valve A type of valve that uses a gate-like element that rises and falls inside the valve to control the flow of water.

Gazebo A roofed but otherwise open garden structure.

Geotextile fabric A material woven from human-made fibers that is buried and provides a barrier between one material and another.

Grading The act of manipulating the earth on a site to give it a specific form or drainage pattern.

Grate An open work of metal bars designed to hold food on a grill.

Gravel Rocks that are typically less than 3 in. in diameter that are free of soil or sand.

Gravity wall A type of retaining wall that utilizes its own bulk, as well as its batter, to hold a hillside in place.

Gray water Plumbing waste that does not include fecal matter.

Grill The part of a barbecue where food is cooked.

Ground fault current interrupter (GFCI) A super-sensitive switch to terminate electrical current in an overload or current leakage condition.

Ground moisture sensor A device to override the automatic controller in an irrigation system when the moisture content of the soil is high.

Grout A mix of cement, sand, small aggregate, and water used to fill the cells of concrete block, locking reinforcing steel into the system; The thin, pasty material pressed in between tile to seal the gaps between pieces.

Guardrail A horizontal railing designed to prevent people from falling off the edge of a high deck or terrace.

Handrail A safety device designed to be grasped while using a stair.

Head The vertical distance from a pump's intake point and its discharge point.

Header A structural member that supports the ends of other structural members at a framed opening.

Header (landscape) A long strip, placed with its top edge at the surface of the soil, that typically separates one land-scaped area from another.

Hearth The floor of a firebox.

Hearth extension A non-combustible area in front of a fireplace.

Heaving The vertical expansion of earth due to freezing.

Hip The sloped ridge at the intersection of two planes of roof.

Hybrid system (drainage) A drainage system that incorporates elements from both open and closed systems.

Jet A type of fountain or spa nozzle that sprays a fairly dense stream of water.

Joint reinforcement A method of strengthening masonry with a framework of welded wires placed in horizontal mortar joints.

Joist A relatively small repetitive horizontal structural member set on edge and spaced evenly.

Junction box A protective enclosure for the connection of electrical wires and/or the attachment of electrical devices.

Lateral Horizontal or generally sideways.

Lateral brace An element that acts to stabilize a structure to resist horizontal forces.

Lattice Open wood screening made of crossing light-weight wooden members.

Leach field A collection of leach lines to distribute the effluent from a septic tank to the soil.

Leach line A perforated pipe within a leach field.

Ledger A horizontal member attached to a wall for the purpose of supporting other structural members such as joists.

Lifts A layer of material placed and spread out on either a sub-base or another layer of the same material, such as one lift of asphalt placed upon a first layer of asphalt.

Light-gauge steel Sheet metal folded into a stiff shape and used for framing.

Line-voltage 110-120 volts, common to residential electrical wiring.

Live load The force imposed on a structure by things other than the structure itself, such as occupants, snow, wind, or earthquake.

Load A force or weight applied to a structure.

Loam Soil consisting primarily of sand, silt, and organic material.

Low-voltage 12-volt electricity that cannot give a fatal shock or start a fire.

Macro-irrigation A system using relatively high pressure usually to irrigate a relatively large area.

Main drain (pool) The primary drain for the entire pool, located at its deepest point.

Manifold A large-diameter pipe for the distribution of water to valves in an irrigation system.

Masonry Bricks, stones, or concrete units bonded together, usually with mortar.

Mastic A thin adhesive material used to secure tile to another surface, such as concrete.

Microclimate A localized climate affected by vegetation, slope, proximity to water or other special conditions.

Micro-irrigation A system using relatively low pressure usually to irrigate a relatively small area.

Miter A butted joint made by bisecting the angle between two intersecting pieces of material.

Mowstrip A 6 to 12 in.-wide band of concrete or other rigid material typically found adjacent to lawn areas that separates that area from other areas.

Natural gas A hydrocarbon gas piped underground as a utility for use as a cooking or heating fuel.

Nosing A cantilevered edge of a stair tread.

Nozzle The part of a fountain spray feature or sprinkler head where water is released from the pipe into the air.

Open system (drainage) A drainage system that is open to the sky, using ponds and swales to convey and address stormwater issues.

Parallam Parallel strand lumber (PSL), which is a composite structural member made of short pieces of wood oriented parallel to the length of the member.

Pea gravel A self-compacting gravel composed of small, round rocks, each approximately the size of green peas.

Percolation The filtering of liquid through the voids in soil due to the force of gravity.

Perforated pipe A type of drainage pipe, either flexible or rigid, that has holes in its sides, allowing water to enter the pipe from the soil that surrounds it.

Pergola A long arbor with paired columns and a pathway at the center.

Permeable Referring to the ability for water and/or air to pass through a material.

Photoelectric cell An electrical switch activated by light.

Pier pad See Column footing.

Pin connection A type of structural connection that acts like a hinge, providing no resistance to rotational forces.

Polyethylene (poly) A thermoplastic resin used in the manufacture of flexible black pipe, pond liners, and other components.

Polyvinyl chloride (PVC) A thermoplastic resin used in the manufacture of rigid pipe for irrigation and pool systems.

Pond liner A flexible pond enclosure, typically made of a rubberized material, that contains the water and prohibits it from percolating back into the ground.

Pond shell A rigid pond enclosure that contains the water and prohibits it from percolating back into the ground.

Pool deck The flat, typically paved surface surrounding a pool or spa.

Pop-up head A retractable sprinkler head that sits flush with the ground when not in use.

Post A small column.

Precast concrete Concrete that has been formed and poured under the controlled conditions of a factory.

Preservative treated (P.T.) Wood injected under pressure with chemicals that retard deterioration.

Pressure regulator A device to limit pressure in a gas or water line.

Pressure treated See Preservative treated.

Propane (LP) A hydrocarbon gas compressed as a liquid in storage canisters from which it is released for use as a cooking or heating fuel.

Rafter The principal repetitive structural component of a sloped roof.

Rail A horizontal member, usually structural, in a fence or railing.

Rain shutoff A device to override the automatic controller in an irrigation system when significant precipitation has occurred.

Rebar An abbreviation for the ribbed steel bar, or reinforcement bar, used to reinforce concrete or masonry.

Refractory Resistant to heat.

Replacement field An area of ground reserved for the installation of a future leach field should it become necessary.

Retaining wall A vertical element in the landscape that allows for the greatest amount of vertical elevation change in the smallest amount of horizontal distance.

Ridge The horizontal edge between two planes at the top of a roof.

Ridge beam A structural support at the top of rafters.

Ridge board A non-structural board to which rafters are nailed.

Ring-shank nail A nail with ridges along its shaft, which help prevent its withdrawal once it has been driven.

Rise The vertical change in elevation from one point to another point.

Riser The vertical face of a step, from the rear of a lower tread to the front edge of a higher one.

Root zone The area beneath a plant, approximately one to one-and-a half times the diameter of the drip line, where its most critical roots are located.

Rotational failure A type of retaining wall failure in which the weight of the mass behind the wall presses the vertical stem over, essentially tipping the wall.

Run The horizontal distance, measured in feet and/or inches, from one point to another point.

Sand A granular material that is well drained but possesses poor fertility.

Sand filter A large bed of sand in a septic system designed to filter effluent before it is piped to a leach field.

Sealer A coating that is applied to the surface of wood or other porous material in preparation for paint.

Seeding The broadcasting of small gravel across the top of wet concrete in order to produce an exposed aggregate finish.

Septic tank An underground chamber for the anaerobic digestion of raw sewage.

Setback A legally required distance from a property line or other boundary on which nothing can be built.

Setting bed A thin layer of mortar, adhesive, or sand that lies between a solid surface material and a solid or compacted sub-surface.

Sheathing The structural skin applied to the surface of a wall or roof.

Shed roof A sloped roof composed of one single plane.

Sheet flow Drainage directed evenly across a tilted, flat surface.

Shrink and swell The tendency of soils, typically those with a high amount of clay, to expand and contract due to fluctuations in moisture content.

Skimmer The first line of defense in the filtration system of a pool, spa, or pond, this element removes large debris from the water before it is cycled through the primary filtration element.

Skirt A loose network of board or other material designed to conceal the area under a deck or porch.

Slope A tilting area of a landscape typically describing a non-paved hill. Slope can also be used to describe the condition of a surface, either in a non-technical manner (a steep slope), or in a more descriptive, technical manner (a 3 percent slope).

Smoke chamber A cavity above the firebox and below the flue of a fireplace.

Soaker hose A porous tube used in irrigation that allows water to slowly leak through its sides.

Soffit A horizontal surface at the eave, extending between fascia and wall.

Spacing The distance between repetitive structural elements such as joists and rafters.

Span The horizontal distance between the two supports of a structural member.

Spot elevation A specific elevation taken or mandated at a specific spot on the site.

Sprinkler head A device for spraying water over an area.

Stain A coating applied to change the color of a surface without forming an impervious film.

Stem wall A vertical wall extending up from, and directly connected to, a footing.

Stucco Cement plaster used as a siding material.

Stud A small vertical structural element used in walls, usually a 2x4 or 2x6.

Sub-base The material, usually crushed gravel, that lies above the subgrade and beneath the paving material, footing, or structure being built.

Subgrade The compacted or undisturbed soil beneath sub-base material.

Sub-panel A small electrical distribution center connected by a cable to the main electrical panel.

Subsoil A less fertile layer of soil located just beneath the topsoil.

Sump A subgrade pit, covered on top, that collects water fed by drain pipes for the purpose of allowing water to percolate back into the subsoil.

Swale A small, typically wide depression in the landscape that collects and directs water from a high point to a lower point.

Swim out A submerged ramp-like or bench-like element within a pool that facilitates entering and exiting the pool.

Swing joint A compound pipe connection that allows both vertical and horizontal movement for final positioning of a sprinkler head.

Synthetic stone Thin, irregular tiles of imitation stone made of lightweight concrete, which are designed to adhere to walls as a veneer.

Synthetic wood A weather-resistant composite material made primarily of recycled plastic and wood fiber and used principally for decking.

Thrust block A block that is firmly attached to the floor at the base of a stair to prevent its horizontal movement.

Toenail A method of nailing diagonally through the end of one piece of lumber into another.

Topsoil The top, most fertile layer of soil, generally composed of a mixture of organic matter, clay, sand, and loam, which possesses limited structural capability.

Transformer An electrical device that changes line voltage to low-voltage.

Tread The level part of a step that the foot touches down on.

Trellis A simple open framework or arbor used to support climbing plants.

Trench drain A drainage structure, long and linear in form, that collects water from a sheet flow and directs it into a pipe.

Turnbuckle A device used to increase tension in a wire or cable.

Underlayment (pond) A cushioning layer placed between a flexible pond liner and the soil below.

Uplift A force that works in an upward direction, such as wind pushing up on a roof.

Voltage drop The lowering of voltage along the length of an electrical line.

Walk out A set of submerged steps within a pool that facilitates entering and exiting the pool.

Waterline tile The band of tile embedded in a pool wall and located at the actual full-water mark of the pool.

Water pressure The force of contained water against an area of the container, measured in pounds per square inch (psi).

Water table The level at which groundwater naturally occurs in the soil.

Weep hole A small hole near the base of a retaining wall that allows water trapped behind the wall to escape, thereby reducing the potential for rotational failure.

Weir A wall or dam-like element in a fountain or stream that creates a pool behind it while allowing water to flow over its top.

Welded wire mesh (WWM) A grid made of steel wire that is embedded within concrete designed to provide added strength.

Wire nut A mechanical connector for splicing electrical wires.

Zone (irrigation) A garden area containing plants with similar watering needs and served by a number of similar sprinkler heads or other watering devices.

Zoning regulation A law that controls land use, including restrictions on the height of structures and their distance from property lines.

INDEX